Dark Sky, Black Sea

DARK SKY, BLACK SEA

Aircraft Carrier
Night and All-Weather
Operations

Charles H. Brown

Naval Institute Press

Annapolis, Maryland

Naval Institute Press
291 Wood Road
Annapolis, MD 21402

Library of Congress Cataloging-in-Publication Data
Brown, Charles H., 1929–
    Dark sky, black sea : aircraft carrier night and all-weather operations / Charles H. Brown.
        p.   cm.
    Includes bibliographical references and index.
    ISBN 1-55750-185-8 (alk. paper)
    1. United States. Navy—Aviation—History. 2. United States. Marine Corps—Aviation—History. 3. Aircraft carriers—United States. 4. Night and all-weather operations (Military aeronautics)—United States—History. I. Title.
VG93.B752   1999
359.9'434'0973—dc21   99-28613

Printed in the United States of America on acid-free paper ∞

06   05   04   03   02   01   00   99   9   8   7   6   5   4   3   2

First printing

Contents

Foreword

This excellent history of aircraft carrier night and all-weather operations tells of the problems—both in the air and with operational commanders—that early night fighters faced when they first arrived in the fleet during World War II. After that, the book clearly chronicles the developments in carrier aviation that now allow every Naval Aviator to fly and fight in any conditions. Looking back, the author does not forget the efforts of the pioneers of night carrier operations.

I commanded the first night fighter squadron to go to sea in carriers. Pete Aurand and I took our squadrons, VF(N)-101 and VF(N)-76, the first night fighters, aboard ship almost simultaneously early in 1944. One episode on that first cruise shows how the fleet slowly came around to accepting the night-fighting skills my pilots had learned. After being established as a full squadron in January 1944, my squadron was then split into two detachments, with Swede Kullberg, my detachment officer-in-charge, in the *Intrepid* and me in the *Enterprise*. Aboard the *Enterprise*, we found ourselves not welcome. The axial-deck carriers had the main mission of launching successive day strikes against the Japanese, starting from before dawn until just prior to sunset. A night launch for us night fighters necessitated delaying preparations for the predawn launch the following day. The air group commander of CVG-10, of which my squadron became a member, declared, "We'll treat you just like any other VF."

Thus, our future didn't look bright aboard the "Big E." Then we got a break. The *Enterprise*'s skipper, Captain Gardner, called me to the bridge one night with the news that a search group of the battleships' seaplanes was lost. The task group commander didn't want to call those airplanes on radio for fear of giving away his ships' position to the Japanese. Gardner asked if I could find the seaplanes with the Corsair's radar and lead them home. I replied emphatically, "Yes!" Launched alone, I had to go only about eighty miles toward Saipan, our next objective, to find them flying a lost plane search square in formation with all their lights on. I could still see the task force on my radar's search mode. Signaling the course to the lost search formation with my turtleback lights, I led the seaplanes back home. I will not forget Captain Gardner's greeting after I landed aboard the *Enterprise:* "All your sins have been forgiven." VF(N)-101 started getting night launches and we ended our cruise with five kills of Japanese bombers, four of them on a single night.

That is how we started "all-weather" flying.

Capt. Richard E. Harmer, USN (Ret.)

Preface

The inspiration to write this book developed from my long-standing interest in carrier night operations, beginning in my first squadron, VC-35—then the only night attack squadron in the Pacific Fleet. Through the years, many people with whom I worked and flew added to my knowledge of that naval aviation specialty. They are too numerous to name, but I could not have absorbed the body of experience on which to base these words without their help.

The immediate spark for the effort came from the interest shown in the project by Capt. E. T. Wooldridge, Vice Adm. D. D. Engen, and Vice Adm. G. E. Miller. After scanning my outline for a history of carrier night and all-weather flying, they encouraged me to go ahead with the work. Mr. Scott Belliveau of the Naval Institute Press guided me to a good start. My son, Brad, gave me a push with tips from his experience as a newspaper and television reporter and as an author. My daughter, Kathy, was helpful as my first editor. And without my wife Janet's editing, patience, and appreciation of the project, I would not have finished the book.

Many men gave their time for interviews in the process of writing this history. I owe them a great deal of thanks for their interest and contributions to the work. I am especially grateful to Adm. Maurice Weisner, Rear Adms. R. C. Mandeville and R. S. Owens, Capts. E. T. Wooldridge and J. R. C. Mitchell, and Cdrs. M. I. Fox and S. B. Barnes, who read the manuscript and offered suggestions for improvement before presenting it for publication. Captain Harmer's review of the manuscript and his foreword are much appreciated. Mr. Hill Goodspeed of the Naval Aviation Museum was a big help during my search for photographs. My thanks also goes to Mr. Roy Grosnick and the staff at the Naval Aviation History Center as well as Capt. S. Milliken of the Tailhook Association's magazine *The Hook* for their help during my research.

In sum, there were many helpers along the course to the book's completion. I give my thanks to all.

PROLOGUE

Dark sky, black sea,
Oh Lord, why me!
Anonymous

Aviators have struggled to fly as skillfully at night and in foul weather as they do in clear, sunlit skies, almost from the time the Wright brothers lifted into the air. In the beginning, a flyer judged the attitude of his airplane by looking at the horizon where sky and earth joined. Without seeing a horizon—which can be obscured by night, clouds, or fog—aviators easily lost all perspective of whether their airplanes were right-side-up or upside-down, climbing or diving. It is natural to fly an airplane with visual reference to the horizon. It is not natural to fly and feel an airplane's attitude by watching only a set of flight instruments. Vertigo (loss of balance) occurs, sooner or later, to every pilot flying with reference only to his or her flight instruments. Even today, Naval Aviators first learn to fly an airplane by visual reference to the horizon before learning the skills necessary to fly at night or in the dim grayness of clouds.

Engineers quickly developed basic flight instruments that promised to take the place of a natural horizon. At first there were only an altimeter, a compass, and an airspeed indicator. Soon a turn-and-bank indicator was added, which gave some perspective of whether the airplane was balanced as it turned. Next a gyro-stabilized attitude indicator was added to the pilot's instrument panel. That instrument became the principal reference for flight in conditions when pilots could not see an actual horizon. Pilots began learning to scan rapidly over their instruments and, from their readings, to understand if their airplanes were upright and level when there was no visual horizon: increasing airspeed and decreasing altitude meant the airplane was diving, decreasing airspeed and increasing altitude indicated a climb, and so on. The problem became one of gaining confidence in the information that the flight instruments gave the pilot.

For some individuals, there was another accomplishment added to the fascination of being airborne: flying from a ship steaming over the ocean. The greatest challenge of all was flying at night and in adverse weather from an aircraft carrier. There was a special, personal reward for individuals who conquered the darkness as well as the air and sea. The aviator who flies only in daylight and in clear weather has a much easier task than the night and all-weather aviator. The pilot flying an airplane with reference only to instruments is under constant pressure and only with continual practice is able to safely control his airplane. Learning the special skills that allow night and all-weather flight operations—and beyond that, combat maneuvers in those conditions—was and is the challenge of the fully accomplished Naval Aviator.

Other than meeting the personal challenge of controlling airplanes while blind to the outside world, Navy men worked hard developing the equipment and tactics needed to operate combat aircraft carriers in all conditions, because there is a strong military advantage for those ready to fight at any time, in any weather. Over time, engineers developed radar and infrared sensors to give aviators the ability to see through and fight in clouds, fog, and rain, as well as at night. The specific wartime missions carried out by carrier night flyers have varied from sinking ships in World War II and harassing trucks and trains in Korea and Vietnam to destroying power plants in Iraq. Today, the concept remains the same: attack and destroy the enemy as soon as possible, no matter what the weather.

This book is a history of aircraft carriers and their fighter, bomber, and attack squadrons (the combat units) as the U.S. Navy developed its ability to fly and fight at night and in any weather condition. This history mentions, but does not tell, the story of the World War II composite squadrons and the later carrier antisubmarine and support squadrons that did and do so much of their flying at night. Part of the history is told from my personal vantage point as a Naval Aviator. The story begins at the inception of carrier aviation when there was no night air combat capability and continues throughout the Cold War. The book ends with vignettes about the carriers' night activity during the Persian Gulf War.

Through the years, the challenges of night and all-weather flying from a carrier have been reduced through technological advances in carrier systems and airplanes. Now, in fact, some say, "The night belongs to us," although we know the night belongs only to God. Perhaps this book will help explain how we have achieved the right to claim at least partial ownership.

Chapter 1 **EARLY CARRIER NIGHT
AND ALL-WEATHER OPERATIONS**

On the night of 8 April 1925 four Naval Aviators assigned to Fighting Squadron One (VF-1) took off from Fleet Air Base North Island, California, and headed offshore for the *Langley* (CV-1). The group, piloting TS-1 single-engine fighters, was about to participate in the first carrier operations to be conducted at night. En route, alone in their airplanes, the night flyers carefully watched their primitive instruments: the airspeed indicator, barometric altimeter, magnetic compass, and a gyro-stabilized turn-and-bank indicator that offered the only aircraft attitude reference to the pilots. Because there were no electronic navigation aids, they used dead-reckoning navigation techniques to guide them during the twenty-mile night flight to the *Langley*.

Through the darkness, the pilots soon spotted the ship's masthead light and the dim white electric deck edge lights that she rigged for the operation. With the *Langley* headed into the wind, Lt. J. D. Price, the leader of the flight, turned opposite to the *Langley*'s course and flew about one-tenth of a mile abeam the ship while descending to an altitude eighty feet above the water, twenty to thirty feet above the carrier's flight deck. Flying at a speed of fifty knots, he checked the position of the ship's masthead light. There is no action in flying an airplane that is more dangerous than trying to watch an airplane's instruments to keep a safe flying attitude and altitude while attempting to see something else outside the airplane at night. Price and his cohorts were the first men to try this technique at sea.

When abeam the *Langley*'s masthead light, Price commenced a turn into the "groove," the segment of the landing pattern flown just before reaching the ship's stern, and began looking through the darkness for the landing signal officer's signals. Price finally sighted the LSO's illuminated wands and followed his signals. When Lieutenant Price was about ten feet above the ramp of the flight deck, the LSO signaled him to cut power and Price completed the first successful night landing on an aircraft carrier.

Lts. D. L. Conley, A. W. Gorton, and R. D. Lyon followed Price aboard the *Langley* that night. Gorton recalls taking off just after dark, circling overhead for a while, and then making four landings. Once over the little square of lights that he could see in the dark, he remembers thinking to himself, there was no turning back.[1]

One can imagine the wind whistling through the canvas-and-wire airplanes' rigging as Price and his cohorts fought the anxiety caused by these initial night landings. At night, there are plenty of ghosts and goblins that still distract carrier night flyers. Those pioneering pilots proved that airplanes could land at night on a moving ship, a giant step in the development of the aircraft carrier's night air combat operations.

■   ■   ■

Preparation for those first night landings aboard an aircraft carrier began soon after VF-1 and VF-2 became the first combat squadrons aboard the *Langley*. The VF-1 pilots performed the first carrier landings in January 1925. Then for two months at least five VF-1 pilots continued practicing carrier landings at night on an area within the North Island airfield that was measured to be the same size as the carrier's deck. At night, tubs of kerosene-soaked rags or lights from automobiles marked the landing area.

Realizing the need for assistance while landing aboard a pitching, yawing, rolling ship, quite soon after beginning carrier flight operations the aviators began using a qualified flyer as a landing signal officer (LSO). Aboard ship, the LSO would station himself near the aft end of the flight deck, monitoring the flight of each airplane attempting a carrier landing. If the airplane was not on the correct speed and altitude needed to land safely, the LSO would signal the pilot of the airplane to make appropriate corrections. The LSO used signal paddles and a modified semaphore flag code to pass his commands to the pilots. At night, on the field and aboard ship, the LSO used battery-powered, lighted wands for his signals. It took many hours of practice to perfect the teamwork between the LSO and pilots that was so necessary to successful carrier night operations.

. . .

Although the first carrier night operation occurred in 1925 soon after squadrons began operating on the *Langley*, experience in night flying did not become a requirement for Naval Aviators until January 1929. At that time each aviator had to fly ten hours at night to become qualified, with at least twenty night landings, though not necessarily aboard a carrier.

During the 1930s, all carrier squadrons practiced night launch and recovery operations. The night flight operations conducted by the combat squadrons, circa 1932, included two phases: navigation and carrier landing training. After four hours of flying solo at night over land, including at least fifteen landings, a pilot started night airways flying. Pilots of the carrier squadrons flew point-to-point navigation flights in their biplane fighters, scouts, and bombers from North Island near San Diego to Los Angeles, Long Beach, or Bakersfield. The length of the Bakersfield mission required one flight during the day, followed by dinner at Bakersfield, after which the pilots would fly the return leg that evening. Returning from night navigation flights, the pilots touched down on North Island's landing area lit by large white floodlights.

Participation in carrier night landing practice began after the pilots were comfortable night flying over land. By the early 1930s, North Island had a more permanent simulated carrier deck, still lit by kerosene flares, on which pilots conducted night field carrier landing practice. Based on their experiences, the squadron commanding officers instructed their pilots in the techniques of carrier night operations. To assist, an LSO attached to either the *Lexington* (CV-2) or the *Saratoga* (CV-3) stayed ashore while the squadron pilots learned the landing pattern and the airplane piloting techniques needed to make a safe carrier landing. During the training syllabus, each pilot performed six landings on the field carrier deck before moving to sea for carrier night landing qualification.

Rear Adm. Dan Gallery remembers that carrier pilots usually performed their four required yearly night landings on a night when a full moon was shining. Gallery's squadron was in the *Langley*, but because she was exceptionally small, the squadrons based in her shifted to bigger carriers, the *Lexington* or the *Saratoga*, for night landing practice.[2]

At sea for a typical carrier night landing qualification mission, the carrier scheduled a full eighteen-plane squadron to launch just after dark. Subsequent to briefings in the ready rooms, pilots manned their waiting airplanes. The flight deck officer signaled takeoff for each aircraft as the pilot taxied into the takeoff spot and turned on his airplane's colored fuselage turtleback light. All takeoffs were deck runs because the carriers did not routinely use their catapults. When airborne, the planes joined in a large echelon as they continued on course forward of the ship. Then, signaling with the turtleback light, the squadron leader moved the unit into a vee of vees, three planes in each vee. After a short flight, the squadron leader shifted his unit into a single-file formation and led them back to the ship. Circling in an orbit over the ship, each pilot descended singly to the downwind leg of the landing pattern.

Only the masthead, deck-edge, and ramp lights were visible from the approaching airplanes. As the fighter or bomber pilot turned into the groove, the LSO's wands came into view. Following the LSO's guidance, the pilot approached the ramp. With the airplane's engine and wings obscuring the ramp lights from the pilot's view, the LSO gave the "cut." Silence, a hard bump, and a sudden stop followed as the plane hit the deck and caught an arresting wire. When all the squadron's airplanes were down safely on deck, another year's night flying practice for that unit ended.[3]

• • •

Practice in carrier night flight operations did not mean that there was serious interest in carrier night combat operations at this time. Pilots' gained proficiency in landing at night only in case a flight remained airborne unexpectedly and had to return to the carrier after dark. However, the night takeoff practice did allow the carriers to conduct predawn launches for surprise attacks on the enemy. Squadrons practiced night bombing rarely and night air-to-air fighting not at all.

Nevertheless, the fleet aviation commanders planned for future use of naval air power. In February 1927, the commander of the U.S. Atlantic Fleet's Battle Fleet Aircraft Squadrons published a memorandum concerning the development of aircraft tactics, which included a description of night air search and attack tactics.[4] Putting those instructions into practice, the first of rare carrier night air combat demonstrations came during the 1929 Fleet Problem. The *Saratoga*'s air group executed a predawn launch of sixty-nine airplanes against the Panama Canal. The Problem's umpires estimated that the air group would have closed the canal when it hit the target shortly after dawn.

The only aviators who routinely flew in rain, fog, and clouds during the 1930s were the Air Mail pilots. Fleet squadrons did not fly extensively during periods of inclement weather, day or night. When the carriers' fighters and bombers received gyro-stabilized attitude indicators to aid the pilots, flying in bad weather became more practical. Still, the carrier aviators generally learned only the basics of straight-and-level flying and climb-out and approach procedures for flying at night and through clouds, and took few opportunities to practice their instrument flying techniques.

Early carrier aviators had no means of talking with other airplanes or with their ships, but it was soon apparent that to be successful in combat night flying, communication among the aircraft and between the carrier and its night-flying aircraft was essential. A major event occurred in December 1928 when the Navy shipped fourteen radio sets to Bombing Squadron Two (VB-2B) in the *Saratoga*. The squadron evaluated radio procedures for single-seat aircraft before the Navy issued the first production order for carrier airplane radio sets in November 1932. By the time the Japanese attacked Pearl Harbor, voice radios were standard in carrier-based fighters and bombers. The multiseat observation and bombing airplanes used for search missions also carried a second, higher-powered radio that had a much longer range than the fighter's voice radio set.

Pilots had only dead-reckoning navigation techniques available to them because there were not yet any radio navigation aids in the carriers. Airborne radio direction finders were still at least a decade away, but near the beginning of World War II, the carriers added a tactical air navigation aid, the YE/YG radio system, to assist returning night flyers. Although its signals were awkward for pilots to follow, the YE/YG remained operational into the 1950s.

There was early development of ship and airplane equipment that could make carrier landings possible without a pilot having visual reference to objects outside his airplane. In June 1933, the Navy issued a contract to the Massachusetts Institute of Technology (MIT) to design blind-landing equipment for carriers. On 1 May 1934, some years after Lt.

Jimmy Doolittle's first instrument landing at an airfield, Lt. Frank Akers made a hooded landing in an OJ-2 aircraft at College Park, Maryland, for the first demonstration of MIT's carrier instrument landing system. Flying under an instrument hood is a task that in many respects is harder than flying in actual instrument conditions. Under the hood, there is absolutely no other flight reference except the airplane's cockpit instruments. Among later flights, Lieutenant Akers completed a hooded flight from Anacostia near Washington, D.C., to nearby College Park.

A little over a year after his first hooded landing at College Park, Akers made the first blind landing on a carrier. He took off hooded from San Diego in an OJ-2, located the *Langley* under way in an unknown position, and landed aboard. Despite MIT's and Akers's excellent pioneering work, however, the Navy did not install an operational instrument landing system in carriers until the 1950s.

Without the development of ship and aircraft air search and detection radars, night air combat would never have been possible. The *Yorktown* (CV-5) received the first ship-based air search radar system and in early 1935 completed testing the system. Six years after the *Yorktown* test of an air search radar, shipyard workers finished installation of the first Radar Plot, the original Combat Information Center (CIC), in the *Hornet* (CV-8). Along with the installation of the *Hornet*'s Radar Plot, the Navy published the directive "Tentative Doctrine for Fighter Direction from Aircraft Carriers." It was basically a guide for controlling fighters in daylight intercepts. However, forward-thinking officers realized that night intercepts of incoming raids could be successful if they modified the doctrine slightly and if trained night fighter pilots were available. By the beginning of World War II, all aircraft carriers had air search radars incorporating a radar-frequency interrogator for aircraft identification— the BT-5 Identification, Friend or Foe (IFF) system.

The night-combat airplanes themselves also needed a sensor allowing the fighters and bombers to "see" at night and in any weather condition. In May 1941, copying a successful Royal Air Force design, MIT engineers started work on an air intercept (AI) radar. By early summer, a prototype Airborne Intercept Model A (AIA) radar was available, capable of detecting airplanes and controlling intercepts at a distance of 3.5 miles from the radar. The AIA radar could also detect surface ships ten miles away.

By August, Chief Naval Aviation Pilot Cecil Kullberg had conducted seventeen test flights using the radar in a twin-engine Lockheed XJO-3, flying from the Naval Aircraft Factory's airfield in Pennsylvania and from the *Lexington.* The Navy was slowly acquiring a night air combat capability, but as the end of 1941 approached, despite the intense night air battles over Europe and the known Japanese interest in night naval operations, the Navy had no night fighters or bombers for the carriers.

Launching in darkness and striking just after daybreak, Japanese carrier pilots destroyed the Pacific Fleet's battleships at Pearl Harbor on 7 December 1941, but failed to find the American carriers. The air groups of the *Lexington* and *Enterprise* (CV-6) flew all that day, searching unsuccessfully for the Japanese fleet. Pilots in the *Enterprise's* air group stayed aloft past sunset and had to use their infrequently practiced night flying skills when Lt. (jg) H. B. Harden waved the entire air group aboard after nightfall.[1]

The Japanese attack served to accelerate the growth of the carriers' night air combat potential as the Navy recognized a need for aviators proficient at night. Soon after Pearl Harbor, night air combat equipment development and night air combat training began in earnest. The Navy established Project Afirm at NAS Quonset Point, Rhode Island, in April 1942. This unit was responsible for developing and testing night fighter equipment and for developing tactics and training officers and men for night fighter squadrons and as night fighter directors.

The first commanding officer was Cdr. W. E. G. Taylor, a veteran aviator who had flown British night fighters against the Germans. He set stringent requirements for future night fighter pilots, demanding that each candidate be an above-average instrument pilot and have over 1,000 hours in the air. There were exceptions: Fresh from flying Wildcats in the *Saratoga*, Lt. Cdr. R. E. "Chick" Harmer had little experience with night flying—not even an instrument card—when he arrived at Project Afirm in September 1943. He learned attitude instrument flying from specialist instructors who taught that skill to student night fighters under the hood in an SNJ, using the "gyro horizon" as the primary instrument.[2]

The U.S. Navy modified its aircraft development plans to provide for incorporation of the air intercept radar in some of the new F4U Corsair fighters. The fleet received the first F4U in October 1942 and, soon after, the Naval Aircraft Factory began mounting the AIA radar in a large dome on an F4U's starboard wing. That modification converted the Corsair day fighter into the first aircraft carrier night fighter, the F4U-2. The bright spot of the delivery of the first Corsair was that Charles Lindbergh delivered it to Quonset Point, along with an inspiring checkout lecture. The neophyte night fighter pilots flew the first of these airplanes to complete Commander Taylor's twenty-nine-week training syllabus in night air combat tactics.

Taylor also took care of the mental and emotional health of his trainees. He had learned that night flying was stressful. The night fighter pilot not only had to keep up a scan of his instruments but also spot a darkened enemy aircraft as he closed for a gun shot. Because his radar could not yet track and automatically fire on the enemy, the pilot could not concentrate on keeping his airplane flying. Taylor knew that success in that dangerous type of flying depended on clear heads and single-minded concentration while in the air. To relax the budding night fighters, Taylor had a party for the aviators every weekend. The "Wing-Dings" began after working hours Saturday morning at 0400, with bad whiskey and passable American champagne.[3]

Besides the AIA developed by MIT, in the spring of 1942 the Navy equipped one TBF and an SBD with the first air-to-surface radars, the Air-to-Surface Type B (ASB). Carrier Air Group Eleven (CVG-11), in the *Saratoga*, received these airplanes. In addition, in October 1942 the aircraft pool at Pearl Harbor received eight TBFs and SBDs with ASB radars. Furthermore, all torpedo bombers eventually had radar altimeters installed. Plans called for the future installation of the BT-5 IFF system in all carrier aircraft to fill the initial electronic suite needed for night and all-weather operations: voice radio, radar, IFF, and radar altimeter.

. . .

By 1943 the night air combat training effort started to bear some fruit, but the first night fighter squadron, VF(N)-75, commanded by Lt. Cdr. Gus Widhelm, deployed to Guadalcanal, not to an aircraft carrier. However, the aircraft carriers, now fully established as the core of the fleet's striking force, and the other ships of the fleet had to counter night threats from Japanese land-based bombers. During the Gilbert Islands campaign in November 1943, Japanese aircraft harassed the task force nightly, but it appeared that the task force's antiaircraft guns could successfully defend the force against the Japanese attacks; consequently, there were no plans for night fighters to augment the antiaircraft gun defense of the ships. Moreover, flight operations at night required the use of lights and radios on the task force's ships. The force commander did not like to radiate those signals that could act as beacons for Japanese submarines and bombers.

Characteristically, despite the lack of the commander's interest, night air combat equipment, and trained night pilots, the aviators were not content to just sit on deck and watch the Japanese night bombers threaten the task force. In the *Enterprise*, Cdr. "Butch" O'Hare, commanding CVG-6, and Lt. Cdr. John Phillips, commanding officer of VT-6, devised tactics for intercepting the Japanese night bombers. Their tactical concept used a radar-equipped TBF to find and guide fighters to a kill. During the same campaign, the air group in the *Bunker Hill* (CV-17) combined an SBD with ASB radar and a Hellcat as a night fighter team, but a night landing crash stopped their night intercept trials.[4]

During the night of 25 November 1943, while supporting the Gilbert Islands operation, the *Enterprise* launched O'Hare and a wingman in F6F-3s without radar along with a radar-equipped TBF-1C in the first attempt to intercept a Japanese bomber at night. Earlier, in March 1943, the Navy had installed the first modern CICs in *Essex* (CV-9)-class carriers. The fighter direction officers (FDOs) then had all their necessary equipment for night air combat operations, but in this instance the FDO and the interceptors failed to acquire the Japanese bombers. The next night, however, the same interceptor team did detect an incoming flight of Japanese "Betty" bombers. O'Hare's group broke up the Betty formation's attack on the task force, but at great expense.

O'Hare and Lt. (jg) W. Skon were in F6Fs. Phillips and his crew were in a TBF-1C with ASB radar. Phillips and O'Hare had planned that Phillips's radar operator, Lt. (jg) H. B. Rand, after making contact with the enemy, would use his radar to direct one or both fighters toward the hostile bombers. In the engagement that actually developed after the TBF made radar contact, however, Phillips quickly spotted one of the Bettys visually—a dark shape with twin, brightly glowing exhaust plumes ahead of him. He did not call the fighters to the attack; while level at 1,200 feet and 190 knots, he shot down the bomber himself with his forward-firing .50-caliber machine guns. During the next five minutes, Phillips saw the dim outline of another twin-engine airplane, which he also fired on and destroyed.

Out of ammunition for his forward-firing machine guns, Phillips called for the flight to rejoin. Sighting an exhaust plume in the darkness directly behind him, Phillips's turret gunner opened fire, thinking the engine glow was from another Japanese bomber or fighter. Butch O'Hare did not join the flight's rendezvous.[5] It is easy to say that Phillips's gunner shot down O'Hare, but some accounts of the fight credit the Japanese with killing O'Hare. In either case, the incident showed that the Navy needed better night fighter equipment and more efficient tactics before the fleet could depend on an airborne defense at night.

. . .

After the Gilbert Islands campaign, the fast carriers returned to Pearl Harbor for ship repairs and crew rest, re-forming in early 1944 with Vice Adm. Marc Mitscher in command of Task Force Fifty-Eight (TF-58), the fast carrier task force. For the first time, night fighter detachments from two squadrons that had trained in Project Afirm went to sea in the carriers. One squadron, VF(N)-101, commanded by Lieutenant Commander Harmer, had the only night Corsairs that flew from carriers during the war. The squadron pilots had learned to control the F4U-2 in carrier operations, an almost unique experience before the introduction of later Corsair versions. The single-crew fighter was fast for its day, armed with six .50-caliber machine guns, and mounted the first AIA radar.

Lt. Cdr. E. P. "Pete" Aurand led the second Navy carrier night fighter squadron, VF(N)-76. The squadron flew F6F Hellcats beginning in the summer of 1943. At first they had F6F-3Es with the APS-4 radar, but the F6F-3Ns with an improved radar, the APS-6, entered the fleet within six months. The Hellcat had almost the same performance and armament as the Corsair, but it was a much better carrier airplane with good cockpit visibility and a slower landing speed.

Arriving in the central Pacific in January 1944, TF-58's air groups cov-

Lt. Cdr. R. E. "Chick" Harmer, *middle of the top row,* commanding officer of VF(N)-101, with the *Enterprise's* detachment of the first night fighter squadron. *National Museum of Naval Aviation, courtesy John W. Kelly*

ered the occupation of Kwajalein and Eniwetok in the Marshall Islands. Harmer's VF(N)-101 had divided into two night fighter detachments, each with three airplanes, nine ensigns, and thirty support personnel, in the *Enterprise* and *Intrepid* (CV-11). Cecil Kullberg, now a lieutenant, took charge of the *Intrepid's* team. VF(N)-76 split into detachments in the *Bunker Hill* and the new *Yorktown* (CV-10). Lt. R. L. Reiserer was in charge of the *Yorktown* detachment.

The carriers of TF-58 were barely ready to perform night operations and their techniques were still unproven, but the Japanese crews flying the land-based Betty torpedo bomber had become more proficient in night harassment and attacks. These long-range bombers could stay airborne and track the task force for hours. They flew at a low altitude, taking advantage of the night and any cloud cover to remain essentially invisible. After locating the task force, the night bomber groups stayed out of gun range, intermittently breaking off a single bomber in a high-speed run at a ship. Sometimes, one bomber would circle around the ships to drop flares, their glow silhouetting the ships as another bomber attacked.

During the campaign, one or more of the four carriers with a night fighter detachment stationed night combat air patrols (CAPs). The FDOs occasionally vectored these fighters from their patrol stations against Japanese bombers. Near Truk on the night of 16 February, a group of Bettys, following their standard tactics, set up a holding pattern about fifteen miles from the center of the task force. Planes of Harmer's and Aurand's night fighter squadrons, sometimes called "Black Chickens," were airborne that night, but their FDOs could not hold contact with the low-flying Bettys. Later that night, Lieutenant Reiserer launched from the *Yorktown* and made radar contact but could not close for a kill, as he failed to spot the Japanese bombers' exhaust flames or dim lights.

A Betty on a thrust out of the Japanese bomber's holding pattern escaped the night fighters and the ship's gunners and launched an aerial torpedo that hit the *Intrepid*. She limped to Hawaii for repairs, taking Kullberg's VF(N)-101 detachment with her. As small consolation, a few nights later, Harmer scored a probable kill of a Betty during a similar engagement. Because of the night fighters' poor success, the carrier skippers and task force admirals did not believe that the night units contributed enough to fleet operations to balance the effort involved in keeping them flying.

Soon after the Japanese attacked and damaged the *Intrepid*, the first American carrier night offensive operations occurred. Lt Cdr W J. "Bill" Martin, commanding VT-10 in the *Enterprise*, was a highly experienced combat leader who also was a strong advocate of night flying and instrument training for all Naval Aviators. VT-10's airplane, the TBM-1C, was designed as a torpedo bomber but also carried bombs. It was a stable airplane, well suited for night work. Its ASB radar's yagi antennas, mounted one under each wing, made the TBM-1C's appearance distinctive. A pilot, radio operator, and gunner filled the three crewmen's seats. The radioman operated the search radar, which was used primarily for locating Japanese ship targets.

Martin and his crews had spent hours of their training time before the Marshall Islands campaign practicing instrument flying and radar bombing. When there were no ships available as simulated enemy ships for night bombing practice, the crews had used rocks jutting from the ocean as radar targets.

The major Japanese fleet anchorage at Truk had been a thorn in the Navy's side since the beginning of the war. Vice Adm. Raymond Spruance, with the larger part of TF-58, including the *Enterprise,* pounded Truk for the first time during the war for two days beginning on 17 February 1944. Earlier, as the task force began the Marshall Islands campaign, Bill Martin had proposed to Vice Admiral Mitscher's staff that he lead a night strike against Truk. Martin and others believed that night strikes would be as effective as day strikes and would suffer fewer losses. The admiral's staff approved Martin's plan, but he had the bad luck to break his arm before executing the strike. Thus, the squadron's executive officer, Lt. V. "Van" Eason, led the first carrier night bombing attack during the night of 17/18 February.

It was a night with a half moon shining. The strike flight of twelve TBF-1Cs made unhurried deck takeoffs with all their lights on bright. After joining, the flight headed westward toward Truk for about forty-five minutes at 500 feet above the water. Reaching a point off Truk, which was visible in the moonlight as a low silhouette against the sky, the flight broke into three divisions.

At 0540, when the divisions were in position to the south, southeast, and northeast of the Japanese anchorages, single bombers began leaving their divisions at one-minute intervals, maintaining a speed of 180 knots at 250 feet above the water. Soon the twelve TBMs had spread on different headings into the Eten and Dublon anchorages in the dark of night. In the steamy South Pacific air, the radar equipment heated the bellies of the aircraft, adding that peculiar smell of hot electrical systems. Picking out distinctive ship radar returns on their ASB scopes, each strike aircraft made four passes at ships in the anchorages, dropping one 500-pound bomb per pass. The crews discovered that there was no air opposition, but the sporadic and generally ineffective Japanese antiaircraft gunfire did cause the loss of one TBM.

When debriefed, the night strike crews claimed that 52 percent of their bombs hit Japanese ships. Post-strike reconnaissance confirmed that the twelve TBFs had sunk or beached thirteen ships.[6] By comparison, in the total of 1,240 sorties flown from the carriers over two days, only twenty-four additional war and merchant ships were sunk during

Cdr. W. I. "Bill" Martin,
commanding officer of
VT-10, stands with his
TBM crew. *U.S. Navy*

daylight hours. The night flyers did far more damage to the enemy for their effort than did the day flyers. In the citation for the Distinguished Service Medal that Martin received for his initiative and leadership, Chief of Naval Operations (CNO) Adm. Ernest King called Martin "a pioneer in this new and exceedingly difficult phase of Naval Aviation, who combined professional skill with outstanding courage, administrative ability, ingenuity and leadership."

. . .

The next major naval operation began on 30 March 1944 and continued for two months as TF-58 with eleven carriers struck the Western Caroline Islands, Hollandia, and Truk. Commander Harmer's detachment of night Corsairs remained part of the *Enterprise*'s air group. By that time Aurand's two detachments had split into three in the *Bunker Hill* and the new *Hornet* (CV-12) and *Lexington* (CV-16). The night fighter detachments participated in the defense of the force as it fended off Japanese night attacks. On 24 April 1944, Harmer made the first confirmed kill of a Japanese raider by a night fighter on a radar-controlled intercept mission. Neither Martin's VT-10 nor any other bombing squadron conducted night attacks during the campaign.

After a short rest, Task Force 58 participated in the Marianas Invasion campaign from 15 June to 10 August 1944. The first fighter sweep from the strike force, launched in the darkest hours before dawn, destroyed one-third of the defending Japanese aircraft. The night fighter detachments participated in the sweep and are credited with destroying at least two Japanese aircraft as they took off from their airfields.

Beginning 19 June, the U.S. and Japanese carrier forces engaged in the Battle of Philippine Sea, commonly known as the "Marianas Turkey Shoot." On the first day of the two-day naval battle, Task Force 58 repelled day-long air attacks from Japanese carriers and shore bases. American fighters and antiaircraft guns destroyed 402 enemy aircraft. On the second day, search aircraft located the retreating Japanese fleet at the maximum range of the American carrier aircraft. Despite the distance to the enemy, Mitscher ordered all carriers to launch strike groups, creating a force of about two hundred carrier aircraft. The strike groups found and hit the enemy in late afternoon, leading to the recovery of that large number of aircraft at night. Mitscher's fame increased when he ordered the carriers, "Turn on the lights!" The VF(N) detachments' aviators helped find and guide stragglers to carriers where the pilots either landed or crashed when they ran out of fuel.

The night recovery operation would have been more efficient if more of the pilots had trained for carrier night landing operations. The training schedules of succeeding air groups changed to reflect that need. Air groups arriving off Okinawa in the spring of 1945 had all been qualified for night carrier landings, if another emergency arose.[7]

The night air combat training unit at NAS Quonset Point had sent four carrier squadrons—VF(N)-101, VF(N)-76, VF(N)-77, and VF(N)-78—to Hawaii by the summer of 1944. When the squadrons arrived at NAS Barbers Point, they split into detachments for deployment in the carriers, sometimes transferring to different carriers. By June 1944, all but one of the seven heavy carrier air groups had night fighter detachments. Although the other detachments had new F6Fs with up-to-date radars,

Enterprise's VF(N)-101 detachment continued flying their old F4U-2s with the AIA radar.

Besides participating in the huge daylight strikes and air-to-air battles off the Marianas, the night fighters continued their night combat patrols, but found few Japanese in the air. The carriers also started using their night fighters on search and attack missions. The *Hornet*'s night strike on Independence Day was typical of those missions. Two F6Fs of VF(N)-76's detachment, piloted by Lts. (jg) John W. Dear and Fred T. Dugan, launched with one 500-pound bomb apiece to locate and strike Japanese ships departing Chi Chi Jima's harbor. After waiting four hours, Dear detected and attacked a Japanese destroyer. Almost simultaneously, three enemy fighters attacked Dugan. More Japanese joined the dogfight and for thirty minutes the two night Hellcats fought at least nine "Rufes," a float version of the Zero.

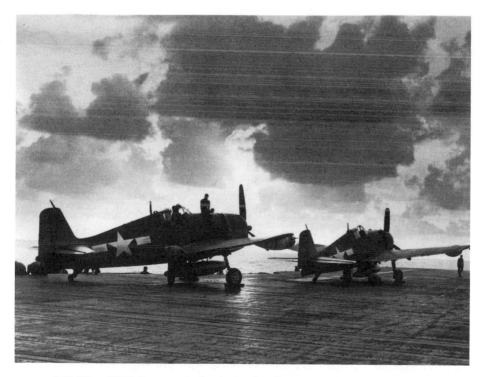

F6F-5Ns of VF(N)-76 ready for dusk takeoff on the *Lexington*. *U.S. Navy*

For the two fighter pilots not only to see so many airplanes but also to maneuver into a firing position and hit the enemy was amazing. A night dogfight was a melee of confused action: A dark blob passes across the F6F's windscreen. The pilot quickly pulls his trigger. Machine guns flash, blinding the pilot for a second. His bullet stream hits the Japanese fighter. Its shadow explodes into a ball of flame. The fighter pilot barely has time to notice that enemy disappear from sight before another target appears.

In the end, Dear and Dugan claimed seven kills, but the Japanese had hit both F6Fs and wounded Dugan. Both pilots returned, not to the *Hornet* but to the *Yorktown*, the closest carrier. Success as a night fighter during World War II required both guts and luck.

. . .

As the fighting progressed, the Night Combat Training Unit at Barbers Point became the second night air combat training organization. Before this unit was redesignated Night Attack and Combat Training Unit, Pacific (NACTUPac) in August 1944, it had absorbed the *Intrepid's* VF(N)-101 detachment. After that time all night combat pilots and crews went through NACTUPac for final training before deployment.

The number of night fighter squadron detachments reached its maximum in August also. Harmer's detachment of the early F4Us had been replaced, and all but one of the night fighter teams had airplanes using the APS-6 radar. That month, night fighter detachments of VF(N)-76, VF(N)-77, and VF(N)-78 were deployed in the *Bunker Hill, Hornet, Lexington, Essex, Yorktown, Enterprise,* and *Intrepid.* Even as their number peaked, however, the night detachments could see the beginning of the end for their units. A Pacific Fleet order dated 10 August 1944 stated that the standard carrier fighter squadron complement would soon include four night Hellcats.[8] The seven night fighter detachments would be absorbed into day fighter squadrons in which their fifty or so night fighter pilots would be a drop in the bucket among the hundreds of day fighter pilots.

As carrier operations became more sophisticated, the concept of a night carrier and air group occurred to a few advanced tacticians in the carrier Navy. After VT-10's Truk strike in February 1944, Commander Martin had recommended to the CNO that at least one night carrier and air group be trained and operated. Kullberg, now a lieutenant commander at NACTUPac, made the same recommendation and met with Rear Adm. Arthur Radford, the operations officer for the commander-in-chief, Pacific Fleet (CinCPacFlt), to discuss the issue. Radford was interested and endorsed the concept of a night carrier and air group.

By the fall of 1943, Lt. Cdr. Turner Caldwell had finished a combat tour in SBDs. Sitting idle on the West Coast, he wrote to his old friend Gus Widhelm at Project Afirm, asking to be ordered to night fighter training. Caldwell became commanding officer of a new VF(N) squadron and was ordered to Vero Beach, Florida, for instrument and night training in the F6F. While there, he also picked the pilots for his new squadron, at which time Lt. Cdr. W. E. "Bill" Henry became his executive officer. When his VF(N)-79 finished training in their F6F-5Ns, Caldwell envisioned a night air group and asked that a TBM unit join his squadron. Caldwell and his squadron with the attached TBM unit left Quonset Point for Hawaii in early 1944.

The concept of sending a night carrier and air group into combat also excited Capt. E. C. Ewen, commanding officer of the *Independence* (CVL-22), who thought the concept could make an interesting experiment. When VF(N)-79, including its TBM unit, arrived in Hawaii, the *Independence* was preparing for another deployment after being in Pearl Harbor for modifications. Captain Ewen and Commander Caldwell met and carried their idea forward to CinCPacFlt. They soon received authority to operate the *Independence* as a night carrier. VF(N)-79, redesignated as CVLG(N)-41, was assigned to that carrier, becoming the first operational night air group.

The original squadron split into VF(N)-41 and VT(N)-41. Besides being the air group's commander, Caldwell acted as the fighter squadron's commanding officer. VF(N)-41 had fourteen F6F-5Ns and five Hellcats without AI radars. The squadron received extra night fighters from VF(N)-78 before sailing. The training in Hawaii gave the night fighters time to get to know the FDOs who had been assigned to the squadron at Charleston.

Lt. W. R. Taylor was the night bomber squadron's commanding officer. VT(N)-41 received twelve TBM-1Ds which had the Air-to-Surface Type D (ASD) radar in a pod fixed on the starboard wing. Some operators could detect ships at eighty miles with a well-tuned ASD radar.

Caldwell's group completed NACTUPac's course and boarded the

Independence for carrier qualifications and deployment when she returned to action in August 1944.[9] Japanese carrier and land-based aviation posed even less of a night threat by then; however, the advocates of night combat flying still saw a need for twenty-four-hour fighter and bomber operations in the fast carrier task force.

Meanwhile, VF(N)-103 had been established at Project Afirm in April 1944. Pacific Fleet plans called for the trained night fighter squadron to split into four carrier detachments. However, when the squadron arrived at Barbers Point in August 1944, it became the nucleus of the second operational night air group, CVG(N)-90. Cdr. Bill Martin reported from the central Pacific to become the air group's commander. Martin's air group replaced Caldwell's unit at NACTUPac. Simultaneously in August 1944, the Navy established another night carrier air group, CVLG(N)-43, at Charlestown, Rhode Island, but within two months, the air group was disestablished and moved its airplanes and aircrews to Project Afirm.

Also in August 1944, Carrier Division 11, Rear Adm. Matthias B. Gardner commanding, with the *Saratoga* and *Ranger* (CV-4), became the first night carrier division. Gardner had been the *Enterprise*'s skipper when O'Hare and Martin experimented with night combat tactics in CVG-6 and CVG-10. Specifically established for night operations, the division stayed in the Hawaiian area as a night training organization. Gardner and the *Saratoga* did get into the battle later, but the *Ranger* was too old to enter the Pacific war.

. . .

When next in action through the month of September 1944, Mitscher's fast carrier striking force provided support for Third Fleet's amphibious landings throughout the south and central Pacific. Ranging farther afield, TF-38 also struck targets in the Philippines from Mindanao to Manila. This should have been a great campaign for the night flyers in the *Independence*'s Night Air Group 41 and the night fighter detachments, all with the latest airplanes and radars, in seven *Essex*-class carriers. However, Mitscher and his staff still believed that night air operations caused too much interference with the conduct of the more-effective day operations, and they limited the work of the night specialists. Naturally, the night flyers believed that their training could be valuable to the fleet and were angry with the decision not to exploit the carriers' new night operations potential. Soon, events proved that their views were correct.

24

Because of concern that one of the "secret" fighter radars would fall into Japanese hands, Mitscher ordered VF(N)-41 to remove its Hellcats' radars for flights over the Philippines, making them impotent for their most important task. Soon after this order was given, however, Lt. Cdr. Bill Henry and Ens. Jack Berkheimer, flying an overwater fighter patrol from the *Independence* on the night of 12 September, intercepted and destroyed a twin-engine "Dinah," an action that broke up a heavy Japanese night attack on the main force. The admiral quickly withdrew his order, and the night fighters immediately reinstalled AI radars on all their Hellcats.

The engagement that night was typical of the night fighter tactics that had matured during the seven months since the VF(N) teams joined the fast carrier task force. Launching for a night mission, a night fighter normally used a rolling takeoff from the deck rather than use the relatively weak catapults then operational in the carriers. The rolling takeoff was much more comfortable for the pilot than a catapult launch. It allowed him to establish his airplane's flying attitude while he rolled down the deck, getting a feel for whether the aircraft was going to fly before he left the flight deck. When launched by catapult, a pilot shot off the deck into the darkness trusting in the ability of the catapult to get the airplane to flying airspeed and altitude while he watched his altimeter, airspeed, and attitude indicators.

The night CAP from the *Independence* or one of the night fighter detachments' carriers remained on station about fifty miles from the center of the task force. After launch, the night fighter pilot reported to an FDO. The night fighter pilot, alone in his F6F or Corsair, spent time orbiting on his assigned patrol station with only his aircraft instruments, radar screen, and radio keeping him company in the blackness. The World War II night fighters had better flight instruments than the SBCs and F3Fs flying before the war; however, the gyroscope platforms for the compass and attitude indicator were still unable to cope with maneuvering flight. If not locked (caged) before entering combat, the attitude indicator failed completely as it tumbled in its gimbals. Even in normal cruise flight conditions, the gyros precessed, causing directional errors in the compass and attitude errors in the attitude indicator. In flight, pilots had to regularly compare the gyrocompass and magnetic compass headings.

Scanning his red-lighted instruments (see figure 1), the pilot focused most of his attention on his attitude indicator and the radar scope. The small, three- to four-inch-square radar screens gave out an eerie green light in the dark cockpits. Target signals or blips appeared as globs of yellow light on the screen as the display's electronic range line swept across the scope. The F6F-3Ns and F6F-5Ns had the APS-6 radar with a five-mile air-to-air range, slightly greater than the earlier APS-4 radar, and a twenty-mile air-to-surface range. The night fighter pilots preferred the APS-6, although it weighed seventy pounds more than the APS-4. Better radar performance offset the loss in flight performance caused by the increased weight. Both radars had B-scopes on which pilots interpreted range and bearing to target blips.

The APS-6 had three scan modes: horizontal mapping, air search, and a targeting presentation. In the air search mode, a ghost blip that indicated the target's altitude relative to the fighter, appeared to the right of the target return. The third mode, a targeting presentation used when closing for a firing run, showed the pilot a target airplane symbol with wings that grew as the range closed. When the airplane symbol's wings spread beyond small vertical lines marked on the scope, the night fighter was within gun firing range.

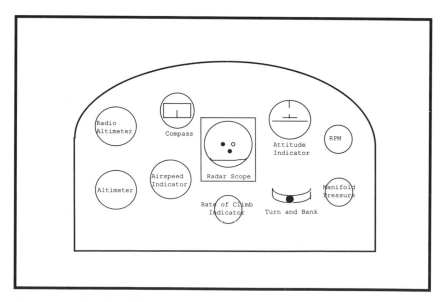

Figure 1. The F6F-5N instrument panel. Note the radar scope.

An intercept started when the carrier's air search radar picked up an incoming Japanese bombing raid. After the force air defense officer selected the night fighters for the initial crack at the enemy raid, an FDO gave one of the pilots an initial vector on which to intercept the enemy. By that time in the war, the FDOs had perfected the art of controlling fighters intercepting enemy raids during the day and at night. The intercept controller attempted to gauge the speed and altitude of the raid and place the fighter at a forward quarter or side position relative to the target.

If the fighter pilot followed his FDO's vectors successfully and detected an enemy aircraft with his AI radar, then the hard work began. The fighter pilots had learned that the radar's gun targeting presentation was unreliable. They preferred to maintain AI radar contact in the air search mode while closing the enemy from his stern until within gun range. A night fighter pilot, having moved to within two or three miles of the enemy using his radar, could usually obtain a visual contact even on the darkest nights. As in daylight, the closer the firing range, the more likely the kill. However, Caldwell said that closing toward another airplane was a bit hairy.[10] Undeterred, night fighter pilots did press their attacks, finding their target visually within their dimly lit gunsights.

The night flyer returning to the carrier after a fight or a quiet patrol turned on his YE/YG to guide him back to the task force. Each carrier had its own assigned YE/YG radio frequency on which she broadcast her identification code and a separate, two-letter signal in each thirty-degree sector around the ship. The fighters had a sector in which to approach the task force. That procedure allowed the force radar operators to identify returning friendly aircraft and made detection of enemy raids easier. The task force changed the YE/YG sector codes frequently to confuse the enemy.

After identifying his carrier's code, the fighter pilot started a circular flight path around the carrier's estimated position. When he heard his sector's two-letter signal, he would turn inbound toward the task force. If the pilot began to lose the correct sector signal, he adjusted his airplane's course to stay in his sector. Immediately over the carrier, the fighter entered a cone of silence and then the pilot heard an abrupt change of the two-letter signal. At that time the pilot commenced a descent to about 1,000 feet above the water. Reversing course, the pilot hoped to spot his carrier dead ahead. The fighter pilot identified his carrier by a unique masthead light that each operating carrier kept on during night aircraft launch and landing operations. The approaching fighter turned alongside its ship to an upwind heading. After passing the carrier's bow, the pilot turned port 180 degrees to enter the downwind leg of the landing pattern.

Because the later carrier flight decks were higher than those of the earlier carriers, the landing pattern altitude had been raised to 150 feet. In addition, because of the faster landing speeds of F6Fs and TBMs, the downwind leg had been placed farther from the carrier than it was in the old F4Bs, to about one-third mile. When abeam the ship's masthead light, the pilot started a 180-degree turn toward the ship's ramp.

Imagine on a dark night trying to see the carrier's single dim red light about six football fields away from you. (The *Enterprise* tried a shielded amber light set on her flight deck, in addition to her masthead light, as an aid for the pilot to judge the correct position abeam. The extra light did not help.) Then try to imagine flying your airplane a scant 100 feet over water that you cannot really see, trusting a single instrument, your altimeter.

Unfortunately, LSOs received little or no practice judging night landings because carrier commanding officers and the admirals did not like to spend time on a steady course into the wind for landing practice. The pilot flew the landing pass entirely on his own until the LSO's battery-powered wands appeared, barely visible in the darkness. The early night LSOs had no lighted vertical or horizontal reference lines on their suits. That made interpreting the LSO's signals even more difficult. Sometime after the Battle of Midway, LSOs began using black lights illuminating fluorescent paddles and horizontal and vertical reference stripes attached to their suits. However, it was not until early in 1945 that the fluorescent suit became standard.[11]

Most pilots did not see the LSO until they entered the last forty-five degrees of turn into the groove, the final part of a carrier landing pass. Centerline lights dimly illuminated the carrier's flight deck for the approaching pilot. After the LSO's "cut" signal, the night aviator had only a bare outline of the narrow landing area for which to aim as he landed. If the night fighter pilot missed an arresting wire, he crashed into the barriers protecting parked airplanes on the forward part of the flight deck.

28

. . .

Trained principally to locate and bomb surface ships and port installations, VT(N)-41 also had learned to hunt submarines, a mission that had become important in the Atlantic Fleet. The hunter-killer task groups in the Atlantic included a CVE with TBMs, and sometimes F4Fs, flying day and night. Operating around-the-clock, these units were successful in reducing the German U-boat threat to Allied shipping. The TBM's ASD radar could detect surfaced boats and sometimes the periscope of a submerged boat. The threat of detection often kept enemy submarines down and away from friendly naval and merchant ships. As used in the hunter-killer escort carrier, two TBMs armed with depth charges, bombs, or rockets, along with guns, worked together. After detecting a submarine, either visually or by radar, the search-and-attack team conducted attacks on surfaced submarines or those seen at shallow depths.

The *Independence*'s night bomber crews were even more frustrated than their fighter pilot cousins. Despite the knowledge that there were few operational Japanese submarines in the war by that time, Task Force 38's staff scheduled VT(N)-41 to fly antisubmarine patrols. Moreover, only 20 percent of those sorties were at night. The squadron flew occasional night search and attack missions looking for Japanese shipping but did not strike any land installations during the September Philippine campaign.

The end of the night fighter squadron detachments came after the Palau and Philippine campaigns. On 24 October 1944, the day fighter squadrons absorbed the night fighter detachments' aircraft, pilots, and maintenance crews. By that date, the night fighters had destroyed fifty-two Japanese airplanes, but of the total kills by the VF(N) detachments, only twenty, less than half, were at night. As members of the day fighter squadrons, the night pilots continued to use their night-fighting skills flying the F6Fs equipped for night fighting, but they flew few night missions, much less got opportunities for night intercepts and kills.

The Philippine invasion was the next operation for the fast carriers. The operation occupied TF-38 for most of October and all of November. The campaign opened as TF-38 struck airfields in northern Luzon, the Manila area, and Formosa for five days. The landings on Leyte occurred on 20 October 1944, followed by two days of intense naval battle during which Third Fleet sank most of the remaining Japanese Navy, either in the Battle of Surigao Strait or during the action off Cape Engano.

Fortunately for the night air combat advocates, the *Independence* remained as a dedicated night carrier operating with the seventeen carriers in the fast carrier force. While the night fighters flew night CAP, the *Independence*'s night bombers flew only antisubmarine patrol or surface-search missions. Two of the search missions were successful.

On the night of 24 October, Ens. J. Dewis, scouting Surigao Strait under a black overcast at about 100 feet above the water, saw and identified Admiral Kurita's battleships entering the strait just before the major surface engagement. Although VT(N)-41 planned a night air attack on the battleships for the night after the surface action, exceptionally poor weather at the carriers prevented launch of the strike aircraft. The squadron had no chance for the coup de grâce on the Japanese fleet as it retired. Also on the night of 24 October, two of VT(N)-41's TBM-1Ds on a scouting mission found the Japanese carrier force near Cape Engano. During the ensuing attack on the Japanese the next day, Task Force 38 sank three carriers, ending any Japanese carrier aviation threat.

. . .

Night air combat training continued at a good pace. By November 1944, Project Afirm had become Night Attack and Combat Training Unit, Atlantic (NACTULant). Because all night crews received their final night combat training at NACTUPac, NACTULant stressed basic instrument flying and radar training. Flight instructors emphasized instrument scan techniques to maintain attitude control at night or in the clouds. The school used SNBs for familiarization flights, during which students learned to interpret the radar scope's range and azimuth orientation, recognize target blips, and interpret a ship or an air target's relative movement. There were a few flights to introduce the F6F or TBM and its radar installation to the student night combat pilot before he transferred to NACTUPac.

NACTUPac trained Cdr. Bill Martin's newly organized CVG(N)-90 from August to October 1944, after which it took its final shape. Lt. Cdr. R. J. McCullough assumed command of the fighter squadron, VF(N)-90, which formed with pilots and men of VF(N)-103, VF(N)-104, and VF(N)-106. McCullough had thirty-two F6F-5Ns or F6F-5Es and two photoreconnaissance airplanes. Some of his night Hellcats were "cannon Hellcats," carrying two .50-caliber machine guns and two 20-millimeter cannon. The cannon's heavier shell provided more killing effect from the short burst that was the only shot the night fighters expected in an engagement.

Lt. R. Kippen from VT-10 led VT(N)-90, the bombing squadron in CVG(N)-90. Before deploying, Martin had that squadron's twenty-one TBM-3Ds modified by removing most of the armor, the top ball turret guns, and the belly guns. There was little or no air opposition from the Japanese at night, so he felt that there was no need for guns and armor. The modification lightened the aircraft structure by about 1,700 pounds, and enabled carriage of more bombs, rockets, flares, or electronic countermeasures (ECM) equipment. Martin's TBM modifications were included in the TBM-3N. The TBM-3D carried the fighter's APS-4 radar in a pod fixed to the starboard wing, but it was tuned to detect surface rather than air targets. Martin had picked three ECM officers for VT(N)-90: Lt. Cdr. Bill Chase, Lt. Cdr. Harry Loomis, and Warrant Officer Bud Jenks. These men guided the installation of newly designed ECM gear in ten of VT(N)-90's TBMs. The TBM was a versatile airplane, slowly inheriting the task of providing passive and active ECM missions, adding to its bombing and antisubmarine warfare (ASW) roles.

Night Air Group 90 embarked in the *Enterprise* during the ship's Hawaiian rest period. The *Enterprise* became the second night carrier when she returned to the fight in January 1945. At the same time, Rear Admiral Gardner reported to the commander of TF-38. He became commander of Task Group 38.5, which included the *Enterprise* and *Independence*, the latter still with CVLG(N)-41. The two carriers and assigned screening destroyers operated as the night task group in the fast carrier task force.

The group's first battle came during the Luzon invasion in January 1945. While one task group of TF-38 covered the landings, the other fast carriers, including the night task group, raided Japanese shipping and shore installations from the coast of Indo-China to as far north as the Ryukyu Islands. For three weeks, the carriers struck airfields on Luzon and Formosa and shipping and shore installations in the South China Sea, Formosa, Hong Kong, and the Ryukyu Islands chain. Inclement weather, common during the winter monsoon and typhoon season, lasted throughout the month.

The fighters of VF(N)-41 and VF(N)-90 flew night patrols in defense of the whole task force starting at dusk and ending at dawn. By then the *Enterprise* was using her catapults as the primary method for launching airplanes at night. There were not many rolling takeoffs, because it was easier to launch the few airplanes needed for night fighter patrols or search-and-attack missions by catapult rather than respotting the flight deck's parking area.

The night fighter patrols usually consisted of a single F6F-5N controlled by the fighter direction officers. Even with the YE/YG operating, night fighters returning to the ship remained under the control of the FDOs. When within fifty miles of the *Enterprise*, the night Hellcats received directions to approach the ship and commence a descent. After getting under the clouds, if there were any, the pilot had to locate the *Enterprise*, which, because of the FDO's directions, was usually straight in front of his fighter's nose. The *Enterprise*, as the night carrier, had a blinking red masthead light to distinguish her from the other carriers and escort ships in the task groups. The night Hellcats and TBMs had radar altimeters, but the instrument was unreliable, increasing the pilots' risk during the night paddles passes flown by the night flyers in both CVG(N)-90 and CVG(N)-41. Tiny flight deck-edge and centerline lights, visible only to pilots in the groove, offered the sole reference for landing.

When not on alert or in the air, the VF(N)-90 pilots spent most of their time in their ready room. The ready room was under the flight deck, forward on the port side, with an entrance to the flight deck catwalk so that the pilots could get to their airplanes rapidly. The squadron's pilots spent most of their free time trying to sleep, but General Quarters alarms warning of a Japanese attack continually interrupted their rest.[12]

While the task groups struck Japanese installations in Formosa during January, Commander Martin's bombers participated in at least one night strike on Keelung, Taipei's harbor. Eight TBMs, led by Martin, sank four ships but lost three airplanes to antiaircraft fire. Martin's tactics were the same as used in the first night strike at Truk almost a year before. Throughout this campaign the bombers of VT(N)-41 in the *Independence* flew continuous antisubmarine patrols and a few day strikes; the squadron did no night bombing. They, as well as VT(N)-90, occasionally dropped antiradar chaff, protecting day strike groups from Japanese radar-aimed guns.

32

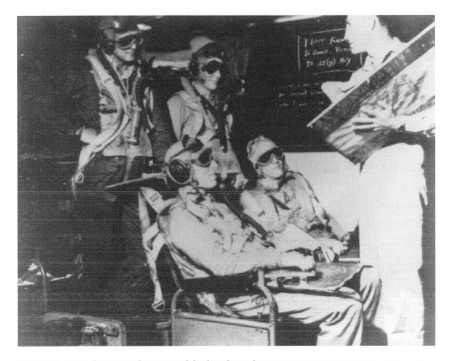

CVG(N)-90 pilots on alert in red-lighted ready room. *National Museum of Naval Aviation, courtesy Bill Darr*

• • •

After the Luzon campaign, the *Independence* ended her long deployment and her night carrier role by going to Hawaii for crew rest and repairs. She had fought from the Western Carolines operation through the last strikes in the Philippines. When she returned for the end of the war, she was a day carrier. CVLG(N)-41, while in the *Independence* from September 1944 until January 1945, lost six F6Fs and pilots, plus two TBMs and crews, to enemy action. The air group also lost nineteen F6Fs and six TBMs, but only two pilots, in operational accidents. More than half of the air group's original airplanes had to be replaced because of accidents. Air group pilots and crews counted twenty-seven Japanese aircraft destroyed at night within the total of forty-six enemy aircraft kills. There were two night aces in the air group: Lieutenant Commander Henry, high with six-and-a-half kills at night, and Ensign Berkheimer.

In December 1944 when the *Independence* departed the forward area, Turner Caldwell left his post as commander of Night Air Group 41 to lead NACTUPac, bringing his wealth of night combat experience with him. By then, NACTUPac had separate training units: a Night Fighter Unit and a Night Torpedo Group. Each syllabus had about sixty-five hours per student in ground school, including discussions with returning fleet pilots about current combat tactics. Each trainee flew fifteen to twenty hours per month during his stay, learning instrument flight and radar operation techniques. Perhaps 30 percent of those hours were at night. The flight training finished with completion of eight carrier night landings, usually on the *Ranger*. The night combat school trained 340 night pilots during the war and also worked closely with the adjoining Air Radar Operators School training air controllers in intercept techniques.

Night Air Group 53 had started night training at NACTUPac in September 1944, just as Martin's crews finished their courses. In late January 1945, the *Saratoga*, with CVG(N)-53 aboard, joined the *Enterprise* as a night carrier in Gardner's TG-58.5 night task group. CVG(N)-53, under the command of Cdr. J. D. Nelson, consisted of VF(N)-53, flying F6F-5Ns; VF-53, a day fighter squadron; and VT(N)-53 in the latest TBM-3Es. The TBM-3E did not have belly guns and so was lighter than earlier versions of the TBM, except for those modified by Commander Martin. The night bomber carried the same APS-4 radar mounted on a rack on the starboard wing as on Martin's TBM-3Ds.

For two days starting on 15 February, Gardner's night task group participated in strikes against the heart of Japan, including airfields west of Tokyo and military facilities in the Ryukyus. These were covering strikes for the impending landings on Iwo Jima. The night flyers also provided night fighter patrols, antisubmarine patrols, and night surface search missions. There was a frontal system between the task force and the Japanese targets that made even day strikes an experience in all-weather flying.

The first carrier strike into the Tokyo area launched before dawn and completed the rendezvous of sixteen F6F-5s under a 200-foot overcast. All pilots managed to stay in a tight echelon formation as they climbed through at least 12,000 feet of cloud cover before breaking into the clear over Japan. On a later strike, flying through the cloud layers induced vertigo in some pilots as they tried to stay with their section or division

Lt. Cdr. W. E. "Bill" Henry, the Navy's leading night ace. *U.S. Navy*

leaders. One reported having to drop out of formation. Because his attitude indicator had already tumbled in the turbulence, he began a swift
scan of his flight instruments—needle, ball, airspeed, altimeter . . . needle, ball, airspeed, altimeter . . . needle, ball, airspeed, altimeter—rapidly
trying to keep up with his airplane. He lost control, falling out of the
clouds, but righted his airplane at 2,000 feet. Recovering from the
plunge, he managed to climb through the clouds by himself and rejoin
the strike.[13]

Expecting tough opposition to the landing on Iwo Jima planned for
19 February 1945, the Marines requested more air support. In response,
the *Saratoga* shifted out of the night group on the 17th to become part of
the direct air support unit. The *Saratoga* operated her night and day
squadrons continuously until put out of action by kamikazes on 21 February. After the damage to the *Saratoga*, the *Enterprise*'s crew continued
the pace of a twenty-four-hour flight schedule for seven days and nights
until 28 February, providing close and direct support to the Marines
ashore. The night bombing done by the pilots was against Japanese units

35

or facilities behind the front lines, for there were no procedures for conducting night close air support (CAS). On 12 March, the Marines opened an airfield on Iwo Jima, relieving the fast carriers.

The dust had hardly settled on Iwo Jima when the Okinawa campaign started. From the middle of March until the third week in June, the fast carrier task force operated in a sixty-square-mile area northeast of Okinawa. From that position, it neutralized Japanese airfields in Kyushu, provided CAS to the Marine and Army units ashore, and provided fighters and antiaircraft guns for defense of the fleet and troops ashore. When the Okinawa campaign began, the night task group had only one carrier, the *Enterprise*. To better distribute the night fighter capability through the task force, the *Enterprise's* fighter squadron, VF(N)-90, received orders to send fourteen of its pilots to day squadrons in other carriers. Those pilots joined the trained night fighters from the former night squadron detachments now assigned to the day squadrons in all the task force's CVs, CVLs, and a few CVEs. The fighter squadrons had four to six F6F-3/5Ns assigned that the night flyers used for fighter or bombing missions.

Throughout April, VF(N)-90 and the other squadrons' night fighter pilots chased Japanese intruders around the night sky. Often the night intercepts lasted between one and two hours before the fighter director and the Hellcat pilot could catch and destroy the enemy. Successful night intercepts required much patience while suffering many near-misses. Despite the difficulties, in April, Lt. D. Umphres of VF-83, Lt. R. J. Humphrey of VF-17, and Lt. (jg) J. Orth of VF-9 became night aces. Although in day squadrons, they flew night Hellcats intercepting night Japanese bomber raids.

The night bombers of VT(N)-90 conducted several night missions attacking the Kyushu airfields from which Japanese fighters and bombers flew. There were some, but not many, reported kills of Japanese airplanes on the ground. However, the night raids kept the fields closed.

Lt. (jg) William B. Balden was a TBM pilot, logging 124 carrier landings during the last years of the war. Only sixteen landings were at night, although he was a member of VT(N)-90 for half of that time. Most of his night bombing missions were on nights that seemed to him to be really black, usually with a cloud ceiling of 300 to 500 feet. Often, sitting in their TBM cockpits, the pilots hoped that the commanding officer would recognize it was too damn bad to fly. But they would inevitably get the "throttle up" signal and be catapulted off the deck at 65 knots. Each pilot knew that if he did not fly his instruments, into the drink he would go. They all felt better once they had leveled out with wheels and flaps up.

A "heckler" mission generally lasted four hours—two hours to the target, then two hours back. On returning to the *Enterprise*, there was a problem of finding what to land on. Balden remembered that it was easy to detect the fleet with radar, but that upon arrival all that would be visible was the white wakes of the ships on the water. The only light kept radiating on the darkened ships was one 360-degree red light on the *Enterprise*. The pilots' problem was finding this red light.

Once the pilots located the carrier, they would pass alongside going the same direction as the ship at 300 feet. This altitude and heading were maintained for one minute. Then came a ninety-degree left turn for one-half minute, followed by another ninety-degree left turn to head back toward the carrier while gradually letting down to 150 feet. When the airplane was alongside the carrier, it would be picked up by radar and the pilot instructed to start a third turn onto the final approach. This was a gradual descending turn getting closer and closer to the water. At this point, the pilots became very tense. Balden said his knees would often tremble with his feet on the rudder pedals. When behind the ship, the deck lights would be turned on and he could soon pick out the LSO—a great relief for the pilot, for the LSO would tell them what to do on the approach. After all of this there was one thing night pilots didn't want—a wave off![14]

The TBMs' crewmen had a different view. Joe Hranek, a radioman, recalled that the catapult shot was a real thrill, especially if everything wasn't tied down. Although crewmen carefully checked their equipment to make sure everything was secure, on one of Hranek's catapult shots a scope cover came loose. As the plane surged forward, the cover shot backward and skinned his forehead. During flight, radarmen often glued their eyes to the windows when not using the radar. On landings they watched the water coming up and recorded the altitude, but could never be sure where they were until the pilot cut the engine and the deck lights suddenly appeared. All in all, it was sheer terror for the crewmen.[15]

By the middle of May 1945, with all VF(N)-90's pilots aboard, the *Enterprise* and her air group conducted an unusually successful night operation. About 2200 on 12 May, the "Big E" launched sixteen TBMs to strike Sasebo and Nagasaki harbor installations and shipping. The TBMs penetrated the harbors' defenses, sank several small merchant ships, and damaged port facilities. To complete their one-two punch against the enemy, about 0230 the *Enterprise* launched a night fighter sweep of F6Fs to cover the return of the TBMs. Japanese fighters followed the TBMs right into the fighters' trap. There was a small night "turkey shoot." The night Hellcats destroyed eleven Japanese aircraft with no losses. Lt. Owen D. Young had the highest score that night as he bagged a "Tony" plus three and a half Rufes.

Two days later, between midnight and dawn on 14 May, VF(N)-90 fighters killed three enemy aircraft while protecting the task force. Then suddenly the Big E's luck ended. A little after the sun rose, a "Zeke" kamikaze knocked her out of war when it hit the *Enterprise*'s forward elevator. After three and a half years of almost continuous combat operations featuring night units, she deserved a better end to her career.

In actions during the previous four months, CVG(N)-90 had made 1,340 carrier night landings, about twelve sorties per night. The fighters shot down thirty-one Japanese airplanes, while the TBM crews destroyed five in self-defense. Lt. C. E. Henderson III, a TBM pilot in VT-10 and VT(N)-90, encountered Japanese fighters in eight fights. The lieutenant had three confirmed and one probable kill. Two of Henderson's kills were at night. The air group claimed 115 Japanese aircraft destroyed or damaged during strikes on their airfields. They sank two and damaged thirty ships.

• • •

V-E Day on 8 May 1945 ended the European war, but not the Pacific conflict. Task Force 38 started the final raids on Japan in July after the Marines and Army secured Okinawa. As the war drew to a close, of the nine CVs in TF-38 only one was a night carrier: the *Bon Homme Richard* (CV-31). Her air group, CVG(N)-91, included VF(N)-91 and VT(N)-91. The *Kula Gulf* (CVE-108) also operated as a night escort carrier with CVEG(N)-63, the last night air group.

VF(N)-91 claimed nine kills in July and August, including one probable and five definite kills in just forty minutes on 13 August. Three Hellcats from the squadron were on a night patrol off Honshu working with a destroyer's fighter direction officer. Vectored initially toward the coast, the fighters detected two unidentified airplanes below them. After destroying those two airplanes, the fighters received a new vector from their FDO. After seeing the third Japanese airplane go down in flames, the flight made another contact on their own and made their last kills.

For a little over a month, the task force executed strikes on airfields, warships, merchant shipping, naval bases, and military installations from Kyushu to Hokkaido. Then soon after Army Air Corps B-29s dropped two atomic bombs on two Japanese cities in early August 1945, the Japanese emperor accepted unconditional surrender terms.[16] Admiral Halsey stopped TF-38 strikes with a message to terminate the war on 15 August 1945.

Night air combat operations had intensified during the war, but not in proportion to the increase in total carrier operations. In the last quarter of the war there were only four night fighter squadrons operating on night CVs, comprising only about 7 percent of the carriers' fighter force. Through one and a half years of operations, night fighter units engaged 115 and destroyed about 100 Japanese airplanes. Half of the kills occurred at night. The top scoring night fighter squadrons were VF(N)-41 and VF(N)-90. There were at least four night aces in Hellcats. The night squadrons lost only two airplanes in combat, an excellent exchange ratio of 50:1.

Of the forty VB and VT bombing squadrons serving on the carriers, few were night bombing units. Only in the last quarter of the war did the VT(N) squadrons appear, maintaining only an average of one night bombing unit in the fast carriers from October 1944 to the end of the war. Because the carrier air groups had destroyed the major part of the Japanese fleet before the night bombing squadrons began full operations, the VT(N)s had little opportunity to destroy more than a small percentage of Japanese shipping. However, while active, the night bombers kept the enemy on the defensive all through the night. Moreover, the VT(N)s were multimission squadrons that, besides executing night bombing strikes, provided antisubmarine patrols and electronic countermeasures support to the fleet.

On balance, in spite of the dedication and sacrifice of the night flyers in the last years of the Pacific war, their results were minor in comparison to the successes of the fast carrier task force's daylight strikes and air engagements. The number of night flyers remained small because there was no real need to exploit the surprise and concealment that night operations offer to skilled combat aviators. Nevertheless, the World War II night flyers proved that aircraft carriers and their air groups could fly and fight at night, leading the way toward a stronger night and all-weather air combat capability in the Navy.

Chapter 3

POSTWAR TRANSITION

When World War II ended, the people of the United States forgot the need to keep an armed force in readiness. Despite President Truman's decision to contain the expanding Soviet Union, the United States's military power declined continuously through the late 1940s. Before the end of 1945, a severe drawdown of naval aviation had begun, forcing the number of experienced night and all-weather aircrews to a minimum. The Navy showed some interest in the carriers' night air combat potential by retaining the NACTUs, which had performed so well during World War II, but it reduced the number of airplanes in the units from nearly five hundred night fighters and bombers to about seventy by July 1946.

In November 1946, after trying another designation for the NACTUs, the Navy renamed the units VC(N)-1 and VC(N)-2 and began using the units for test and development, as well as for maintaining the night and instrument flying skills of a cadre of carrier aviators. The high experience level of the squadrons' pilots made them ideal for testing new airplanes and equipment for night air combat tasks. The units tested the F7F-4N Tigercat, a new night fighter, and the F8F-1N Bearcat, a modification of the postwar fleet day fighter.

Neither unit favorably endorsed the Tigercat because it was an unsatisfactory aircraft for carrier operations. It had poor roll response, floated on landing, and was too big for easy hangar deck handling. However, the Marine Corps successfully adopted the airplane for its night fighter squadrons that operated from land bases. The night Bearcat was worse than the Tigercat. The APS-19 radome mounted on the Bearcat's star-

41

board wing reduced the airplane's maximum airspeed by almost a hundred knots. With that performance, the night Bearcat could not beat the old night Hellcat in a dogfight—clearly an unsatisfactory outcome. VC(N)-2 tested the F4U-5N during 1947 and 1948. That version of the Corsair was an excellent carrier airplane and replaced the F6F-5N as the operational night fighter-bomber before the Korean War erupted.

An airplane that lasted a long time in the Navy, the AD Skyraider, was first assigned to fleet attack squadrons in December 1946. That single-seat aircraft was the backbone of attack aviation, including night attack, for about fifteen years. By 1948, the first night AD arrived at VC(N)-2. The AD-1Q was a three-seat, APS-19-equipped, night attack, ASW, and ECM airplane. Its bomb load was somewhat less than the 12,000 pounds carried by the single-seat ADs, but it was still a formidable attack aircraft. The radar operator and the ECM operator sat in the fuselage of the AD, behind and below the pilot. The AD-1Qs and the newer AD-2Ns quickly replaced the World War II TBM-3E/Ns as night attack airplanes and submarine killers.

The Naval Air Test Center at Patuxent River, Maryland, and VC(N)-2 also tested searchlights, already used for submarine hunting and surface search, as an aid to night fighters. Carried on an F6F-5N's port wing, the Navy found the lights ineffective.

In parallel with the airplane and airplane system advances tested in the VC(N)s, the *Essex*-class carriers began a series of major modifications that allowed them to operate heavier aircraft and store more aviation fuel and ordnance. Moreover, there were improvements in air search radars and other electronic system advances, including the little-noticed but important Tactical Air Navigation (TACAN) system, which started development in 1948. Lastly, in the late 1940s the Naval Air Test Center tested a version of an instrument carrier landing system, the first since the 1920s, but the Navy did not incorporate the system in any carrier.

By March 1948, 80 percent of the Navy's total night combat aircraft inventory—about 180 F6F-5Ns and TBM-3E/Ns—belonged to VC(N)-1 or VC(N)-2. With those airplanes, on 1 August 1948 VC(N)-1 and VC(N)-2 were redesignated as Fleet All-Weather Training Units, Pacific and Atlantic (FAWTUPac/FAWTULant), still based at NAS Barbers Point, NAS Miami, and NAS Key West. This was not just a name change but foretold a more important reorganization of carrier night fighter and attack operational units. The men in both FAWTUs became the leaders and instructors in the complicated art of night and all-weather combat flying. Cdr. Bill Martin continued in a leadership position in the field by joining FAWTUPac as executive officer to Capt. Paul H. Ramsey.[1]

In spite of the general disregard of night and all-weather operations, at least one air group, Carrier Air Group 21 under Cdr. Harvey Lanham, had a full squadron of F6F-5N night fighters. This squadron, VF-212, was commanded by Lt. Cdr. C. P. "Charlie" Muckenthaler. He had been a night fighter pilot in World War II and was initially the only pilot in the squadron with any night experience. He sent a few of the VF-212 pilots to FAWTUPac for the instrument and all-weather flight training course, where they also learned the basics of night air combat. Muckenthaler added a graduate phase to the new night fighters' training, at times introducing a high degree of realism by ending a usually straightforward night intercept practice flight by turning out the lights on his Hellcat. That produced the same spine-tingling, turning, now-I-see-him-on-the-radar-now-I-don't night fight that combat pilots had engaged in during the war. Luckily, there were no midair collisions.[2]

. . .

By the late 1940s, it was apparent that the Soviet Union was building a submarine force larger than the Germans' World War II U-boat fleet, threatening the United States and NATO's control of the sea. To oppose that threat, the antisubmarine (VS) squadrons continued hazardous night and all-weather operations from the small CVEs as they had done to destroy the German boats. Some men with that experience claim the job was insane, but the challenge kept them at the task.

The Navy had discovered in the latter part of World War II that the search-radar-equipped TBMs could also detect low-flying kamikazes. Some of the ASW aircrews, therefore, began to fly what became known as airborne early warning (AEW) missions. The Navy established the first AEW units, VAW-1 and VAW-2, in the Atlantic and Pacific Fleets in 1948. The ASW and AEW units flew at night and in any weather condition, keeping a constant search for aircraft and submarines threatening the task force in which they operated.

Some say that without the need to protect ASW and AEW aircraft from enemy aircraft and surface ships at night and in adverse weather conditions, the night air combat mission would have been scrapped in the late 1940s. It was not until after the establishment of the VAW squadrons that, in March 1948, the first postwar night fighter and attack teams deployed with carrier air groups from VC(N)-1's Detachment One at San Diego. VC(N)-2 also sent night air combat teams to sea from its East Coast air stations. When those units were reassigned, the pilots and aircrews in the FAWTUs also made several carrier deployments during 1948's closing months, continuing the VC(N) squadrons' task. A FAWTU night fighter detachment (VFN) boarded the *Tarawa* (CV-40) and cruised around the world with her, while other night teams deployed in the *Boxer* (CV-21) and *Princeton* (CV-37).

Realizing the difficulties that the FAWTUs faced in performing both training and operational missions, the Navy formed night fighter composite squadrons to continue providing night and all-weather air defense for the operating task forces. FAWTULant left the operational part of the Navy, except for her detachments already at sea, in September 1948 when the Navy established VC-4 at NAS Atlantic City, New Jersey. The last of FAWTULant's detachments returned to Miami in May 1949. FAWTULant became responsible only for basic night air combat training and its assigned test work.

VC-4 assumed responsibility for the final operational training and deployment of Atlantic Fleet carrier night fighter and attack (which were also submarine-killer) detachments. The squadron's first airplanes included forty-four combat airplanes, including F6F-5Ns, TBM-3E/3Ns, and AD-2Qs. By December 1948, VC-4 had trained and sent F6F-5N fighter-bomber teams in the *Philippine Sea* (CV-47) and *Midway* (CVB-41) for deployments with the Sixth Fleet, beginning its history of maintaining night and all-weather combat detachments in the Atlantic Fleet's carriers. At NAS Atlantic City, the squadron conducted flights aimed at providing instrument qualification cards to all its pilots. However, by June 1949 only 58 percent of VC-4 pilots had a standard instrument card.[3] If only half of the Navy's night fighter pilots were instrument qualified, what was the state of night and all-weather flying in the rest of the Navy?

The last FAWTUPac carrier all-weather team participated in the Minor Cold Weather Exercise off Kodiak, Alaska, in February 1949. Then, in May 1949, the unit lost its operational mission when the Navy established VC-4's sister unit, VC-3, on the West Coast. The new squadron's pilots, men, and airplanes transferred from FAWTUPac's Detachment

One. VC-3 moved to NAS Moffett Field, California, in October of that year, continuing operations with only twelve combat airplanes on board. Cdr. Chick Harmer, the World War II night fighter leader, became commanding officer of VC-3 in November 1949. In a serendipitous move, Harmer started a flight training syllabus that emphasized day and night bombing and strafing practice rather than night fighter tactics. Using the target ranges near NAAS El Centro, California, the night fighter-bombers participated in bombing exercises and conducted day and night field carrier landing practice (FCLP) at the El Centro airfield.

Meanwhile, the Navy established another night squadron, VC-33, on 13 May 1949 at NAS Atlantic City. At that time, naval aviation's seagoing antisubmarine capability existed only in the VC or VS squadrons operating in the CVEs. Initially, VC-33's mission was to provide antisubmarine units to deploying Atlantic Fleet CVs and CVBs, but the new unit's ASW mission was short-lived. There are no records of VC-33 deploying an ASW detachment before the Navy changed the squadron's primary task to night attack.[4]

. . .

The Soviets, America's primary threat in the late 1940s, were known to be developing a fleet of turbine- and jet-powered bombers resembling the U.S. Air Force's jet-powered, high-altitude bomber designs. These bombers would be capable of locating and attacking ships at night and in any weather condition. In concept, the threat of Soviet air raids against the carriers was identical to the threat from the Japanese bombers in World War II. To defeat the future high-altitude bomber threat, the carrier fighter squadrons gradually received their first jets, the FH-1 Phantom I and FJ-1 Fury. In 1948, Cdr. Pete Aurand, the former night fighter commanding officer of VF(N)-76, flew an FJ-1 during its initial carrier qualifications in the *Boxer*. Later that year, VF-51 was the first squadron to qualify in a jet carrier airplane, again the Fury, in the Pacific Fleet. Most of the pilots in VF-51 had transferred from the night fighter squadron VF-212 when Carrier Air Group 21 disappeared in a force reduction action not much more than a year after it formed.

While flying propeller-driven airplanes that were efficient at lower altitudes, Naval Aviators could easily conduct their missions underneath most overcasts. With the introduction of jet airplanes, which fly most efficiently at altitudes in or above the cloud tops, carrier pilots were forced to become proficient in instrument flying. In 1949, the *Valley Forge*'s (CV-45) air group commander persuaded the ship's commanding officer to install the first carrier low-frequency radio homer that worked with an automatic direction finder (ADF) in the airplanes. The homer and ADF systems, as they replaced the YE/YG, made night and all-weather approaches to the carriers much easier. By the summer of 1949, Lt. Donald D. Engen, a member of VF-51, had codified the teardrop approach using the ship's homer and ADF systems. Engen's approach pattern (see figure 2) became the carrier's standard instrument approach for jets until after TACAN systems became common.

In September 1949, the Navy began a test of the latest runway visual approach aids at the Landing Aids Experimental Station, a small field at Arcata, California. FAWTUPac's experienced night and all-weather pilots flew F7Fs and ADs in the tests, making approaches and landings under near zero-zero weather conditions. The pilots described the experience of flying in minimum weather conditions as invaluable. Because of his night and instrument flight experience, Lieutenant Engen flew the first jet fighter used during the test. The weather conditions were perfect—in other words, absolutely horrible—but the FAWTUPac instrument training made it possible, with the guidance of an expert ground-controlled approach (GCA) crew, to land in the fog prevalent at the field. Although not carrier operations, flying in those conditions proved that jets flown by a Naval Aviator could operate in the worst weather.[5]

The Navy had begun developing means by which the carriers could take nuclear weapons to sea shortly after the Army Air Force dropped the first nuclear bombs on Japan. The carriers' requirement to deliver nuclear weapons became more obvious as the Soviet nemesis grew. To fulfill that requirement, the carrier Navy needed a new, long-range airplane capable of delivering nuclear weapons at any time in any weather conditions. The pilots and crews of the carriers' nuclear strike airplanes would have to launch, penetrate enemy defenses, find their targets, and accurately drop their weapons before returning to land on their carriers.

Aircraft engineering work begun in 1946 culminated in September 1949 with the delivery of the first AJ-1 Savage to VC-5, the first squadron established as a carrier-based nuclear weapon delivery unit. The first carrier aircraft capable of delivering nuclear weapons, the Savage was also the heaviest carrier airplane of its day. Its three crewmen flew the bomber

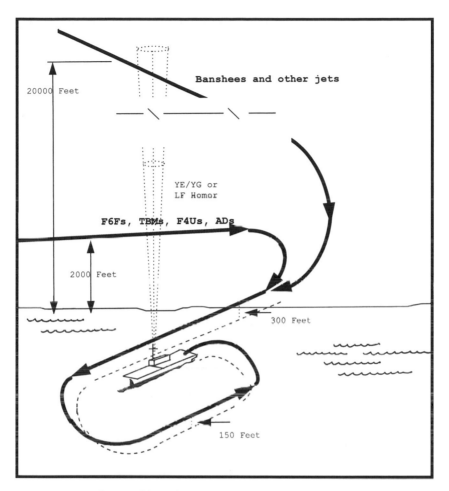

Figure 2. Teardrop and low-altitude approaches to the paddles landing pattern.

and operated the AJ's fine air-to-ground radar and bombing system that made night and all-weather operations possible. The next spring, Capt. J. T. Hayward, commanding officer of VC-5, made the first carrier takeoff and landing of an AJ in the *Coral Sea* (CVB-43), one of a new class of carriers larger than the *Essex* ships.[6]

• • •

Five years past the end of World War II, despite the addition of jet fighters, the inauguration of the all-weather nuclear strike mission, and the establishment of the night all-weather fighter and attack squadrons,

Navy leaders still gave little attention to night air operations in the fleet or to night and instrument flight training. In the summer of 1945, student pilots in the Naval Air Training Command (NATC) received only about ten hours of flight time dedicated to familiarizing them with night flying. After receiving their wings, a few new aviators, or "nuggets," did attend All-Weather Flight School, a thirty-flight-hour instrument course in the SNB, of which about one-quarter of the flight hours were at night.[7]

After reporting to a fleet squadron, new aviators found that the standard combat airplanes—Bearcats, Corsairs, Skyraiders, and the early jets— were ill-equipped for instrument flying. Because of the cockpit design, pilots could not easily scan their basic flight instruments. Partially because of poor equipment, when night fell or clouds appeared, most fighter and attack pilots attended to their collateral duties or broke out the acey-deucey gameboards. Aviators flew about twenty hours a year at night in their type of aircraft, but there was little instrument, night tactics, or night carrier landing practice.[8] Squadron commanding officers did schedule instrument training flights in which pilots paired off to act as safety lookouts for each other. However, time under the hood was rare on those flights, because most carrier pilots considered any time in the air as an opportunity to tail-chase and practice dogfighting. Rarely, selected pilots in day fighter or attack squadrons went to All-Weather Flight School or one of the FAWTUs for instrument training. Upon returning to the operational squadrons, they tried to pass on their new skills to squadron mates. Some nuggets circa 1949 went to short instrument courses taught by instructors from the FAWTUs before reporting to their first squadrons.[9]

There were signs that Naval Aviation was becoming an organization in which all pilots and crews would be able to fly at night and in adverse weather conditions. Yet, when the squadrons stationed in the Northeast wanted to accomplish any serious carrier or weapons training, they avoided the usually bad flying weather at home and always deployed to Jacksonville, Florida, or Guantánamo Bay, Cuba.[10] Furthermore, although the Navy did require all carrier pilots to perform two night landings a year, the carriers' operating periods at sea required to meet that qualification were often canceled because of the austerity of the period.

In January 1950, all of the Navy's carrier night and all-weather air combat capability rested in the FAWTUs, the two night fighter squadrons

(VC-3 and VC-4), and the two nuclear weapons squadrons (VC-5 and VC-6). The Navy established its first unique night conventional (non-nuclear) weapons attack unit when VC-33 became the Atlantic Fleet's night attack squadron in May 1950, taking more TBMs and ADs, officers, and enlisted men from VC-4. That same month, the Navy established VC-35 at NAS North Island as VC-33's sister night attack squadron. VC-35 received her men and equipment, including airplanes, from VC-3. VC-33 and VC-35 shared a mission to provide for the operational training and deployment of VAN detachments for night attack, ASW, and ECM tasks in CVs and CVBs.

In the late 1940s, experienced volunteers had filled the night fighter and attack squadrons, but in January 1950 the Navy reverted to its World War II policy and began assigning some new aviators from NATC directly to a FAWTU or night VC squadron. A VC-4 report stated that the squadron quickly assimilated the inexperienced pilots into the all-weather program. The squadron's leaders believed that while the new aviators might not have been as good as the more experienced pilots, they reached an adequate skill level. Furthermore, VC-4's commanding officer suggested in another 1950 report that all carrier pilots be all-weather qualified as they finish flight training. After that training, the designation "all-weather pilot" would be unnecessary.[11] Putting that recommendation into effect, by the fall of 1950 Naval Aviation required all carrier pilots to qualify for an instrument card showing their ability to fly in instrument flight conditions.

When assigned to a night unit, the pilots, experienced flyers as well as nuggets, went to a FAWTU at NAS Key West or NAS Barbers Point. FAW-TUPac's and FAWTULant's mission in June 1949 included training and indoctrination of VFN, VAN, VAW, and VS pilots in instrument flying, qualification of those pilots for a "Standard" instrument rating; qualification of aircrews in the operational use of radar, armament, and special equipment; providing additional air controller training to selected graduates of the Combat Information Center Officers' School located near NAS Glynco, Georgia, to fully qualify them for the fleet's day and night air controller billets; testing all-weather equipment and techniques; and participating in joint air defense efforts when directed by higher authority.

Both FAWTUs had similar courses designed to fulfill their training mission. The night fighter course took about one hundred hours in the air and five months to complete. It started in the Link trainers, where pilots practiced the old "Charlie" and "Dog" basic instrument flying patterns. Flying these patterns in the Link and in the air taught a night pilot the finer points of three-dimensional flight while he watched and followed only the indications of his flight instruments. The standard instrument panel still included a reliable altimeter, airspeed indicator, turn-and-bank indicator, compass, and a poor attitude indicator.

There were lectures about night vision, because all night intercepts culminated in a visual sighting of a black airplane against a nearly black sky. A successful night fighter pilot had to learn the technique of looking out of the corner of his eye, using the eye's rods (the nerve endings in the eye most responsive to seeing in the dark). FAWTUPac used an F6F-5N cockpit in a black-painted room as a simple but effective night fighter simulator. A light projector showed an airplane's black shadow on the wall in front of the night fighter student seated in the cockpit. The pilot learned to pick out the black-on-black target using visual techniques the instructors taught him.[12]

An experienced fighter pilot with orders to VC-3 or VC-4, passing through a FAWTU, had to fly a lesser-performing airplane than he had been flying in the fleet. He moved from F8Fs, F4Us, or jets to SNBs and F6F-5Ns, but for the new aviator, it meant flying airplanes similar to those he flew at NATC. The FAWTUs' flying course began with fifty hours in SNBs, where the student night pilots learned the APS-6 radar operating techniques while seated at one of several radar consoles in an SNB's cabin. Then the future night fighter pilots had the fun of flying 125 hours, about 70 hours at night, in the Hellcat while practicing night-fighting techniques.

Besides night intercept training, the prospective night fighter pilots also received some radar bombing and rocketry training. Most pilots were concerned about the necessity of closing their eyes when firing rockets because they lost their night vision in the flash of the motor. The night attack pilots and aircrews concentrated on low-altitude laydown night radar bombing to destroy enemy ships, a mission remaining from World War II. There was some emphasis on interpreting radar images of land terrain features and manmade installations. The night attack pilots also learned visual techniques for night dive bombing, rocketry, and strafing attacks. Dive bombing, like the final stages of a night fighter's intercept, required the pilot to split his visual scan from his instruments to the world outside the cockpit. Staring through the blackness trying to see a gray-and-black target on the ground often caused a pilot to fly into the ground or stall and crash his airplane.

The FAWTUs separated the syllabus for all-weather attack aircrewmen sent from VC-33 and VC-35 (the heavy attack squadrons trained their own crewmen) into two phases: basic and advanced. The basic course lasted about six months and consisted of survival, safety, familiarization with aircraft, basic electronics, communications, radar operation, recognition, and ordnance. The advanced course consisted of training in advanced radar image interpretation, dead-reckoning navigation, radar bombing, countermeasures, and meteorology. The fledgling aircrewmen began flying in the SNB. Practicing with the APS-6 and APS-19 radars, they learned the difference between the radar images of sea return, mountains, ships, manmade objects, and clouds, as well as how to track a blip on the scope as his airplane closed for an attack. Then the crewmen and their prospective pilots trained together on navigation, bombing, countermeasures, attack, and intercept flights in the F7F, AD, or TBM.

The night air combat schools did not ignore air controller training. In accordance with the FAWTUs' mission, selected fledgling air controllers advanced from the basic courses at Glenco to either FAWTULant or FAW TUPac. By December 1949, each all-weather unit was training ten low-visibility intercept controllers every six weeks. These men became air controllers in carriers, battleships, cruisers, and destroyers.

. . .

Once the new night combat pilot or crewman finished a FAWTU night fighter or attack course, he reported to one of the four night combat squadrons. Although there were many jet fighter squadrons in the fleet, at the beginning of 1950 there were no jet airplanes in the night combat units. The operational night fighter was the F4U-5N. An improved version of the World War II fighter, it had the same power plant, a larger canopy, a primitive autopilot, good basic instruments, four 20-millimeter cannon, and an APS-19 radar. The worst attribute of the Corsair was its long nose. Carrier landings, especially at night, were always scary in the "Hose Nose"; once lined up with the deck and given a cut signal, the pilot could not see straight ahead.

51

The night attack pilot began flying a three-crew version of the AD equipped with the APS-19 radar. Two crewmen sat in the belly of the airplane operating the radar and ECM equipment. Gaining confidence in each other and developing teamwork became an important part of the attack pilot and crew's training.

The new night pilot began carrier night operations in his new squadron. Carrier night air operating procedures had not changed significantly since World War II. These were the same for all night flyers because both fighter and attack units still flew propeller-driven airplanes with the same sensitivity to altitude, fuel, and icing conditions. The equipment had not changed, either. In 1950, the most common tactical air navigational system at sea was still the YE/YG, although there were some carriers with low frequency homers, operating with airplanes that had the ARN-6 ADF system. Neither did the carriers have a good precision radar system to direct returning airplanes to a landing through clouds, rain, and fog. Flights of night fighters or attack airplanes had to maintain their own separation while approaching the carriers to enter the night landing pattern. For that reason there were few instances of large numbers of aircraft flying at night or in bad weather.

The greatest accomplishment of the carrier night pilot was to meet the challenge of and become proficient at night landings, but nothing had been done to eliminate the threat of a barrier crash if pilots made a small error in the low-altitude, dimly lit, paddles-aided carrier landing approach. Carrier night landings on the straight decks blocked by barriers were extremely dangerous, costing the loss of many airplanes and aircrews.

In comparison with night landings, night air combat techniques for intercepts and bombing were relatively simple because pilots had more clearance between their airplanes and the ground. The night pilot completed his training before deploying with a VFN or VAN detachment by practicing night combat procedures and tactics. After about 150 hours in his fighter or attack airplane, the new night pilot was ready for sea.

• • •

By April 1950, VC-3 and VC-4 owned about seventy night fighters, mostly Corsairs and a few of the new night attack ADs. Because Harmer's training philosophy for VC-3 emphasized night air-to-ground techniques, VC-3 pilots, who flew either fighter or attack aircraft, were ready when the Korean War started. The squadron deployed its first night fighter and attack detachment in the *Boxer* to the Western Pacific during January 1950. From that time until the late 1950s, the carrier air groups included VC teams for all night fighter, night attack, antisubmarine, and ECM missions.

The first VFN team had seven pilots and an LSO qualified to guide night landings. The team's maintenance crew had thirty-eight enlisted men to care for six F6F-5Ns. The night team shared a ready room with one of the day F4U squadrons aboard the carrier. That pairing of a night team and a "host" fighter or attack squadron lasted throughout the existence of the night fighter and attack VC squadrons.

Chapter 4

THE KOREAN WAR YEARS

When the North Korean army stormed into South Korea on 25 June 1950, rapidly forcing the U.S. and South Korean troops into the Pusan area at the tip of the Korean peninsula, the United States began a desperate effort to rebuild its armed forces. That summer as idle carriers returned to join the active fleet and Reserve Naval Aviators and maintenance men filled new air groups, the *Valley Forge* (CV-45) was the only carrier in the Western Pacific ready to help the endangered United States and South Korean troops facing the invading force.

In May 1950, Lt. Cdr. Bill Henry, the World War II night ace, had deployed as officer-in-charge of VFN Detachment C in the *Valley Forge*. Her air group, including Henry's team flying four F6F-5Ns and three AD-2Ns, provided the Navy's share of air support to the ground units for weeks. Because of the shortage of air support for the troops, the VC-3 night team flew daylight missions. Although Rear Adm. E. C. Ewen, the first commander of the Western Pacific's carrier task force, Task Force 77, had been the first night carrier's skipper during World War II and was a strong supporter of night operations, he judged that the night flyers were better used in daylight.[1] Henry had had a similar experience in World War II, when his night fighter squadron flew a high percentage of its missions during daylight hours. The night fighters were easily made available for daylight missions because the Communists had few airplanes and no known night air capability. Commander Harmer's bombing and strafing syllabus proved its worth, but there was little night work as the two VC-3 teams flew day direct support sorties against the North Korean army units into late summer. During those early days of the war, the night detachments also flew day interdiction missions along the Korean roads and railroads, attacking and destroying trucks and locomotives.

By September, the *Philippine Sea* and *Boxer* had joined the *Valley Forge* off the Korean coast. The next operation of the war, a successful amphibious landing at Inchon, forced the North Koreans to retreat from their positions surrounding Pusan and began an extended United Nations drive toward the Yalu River, the Chinese border. The night teams continued flying direct support missions for the ground units and some interdiction sorties during daylight, because the air commanders still did not believe that night missions would contribute significantly to the campaign.

In October 1950, after the *Leyte* (CV-32), the first Atlantic Fleet carrier sent to Korea, arrived in the Sea of Japan with VC-4 and VC-33 detachments, the night teams of the four air groups began to fly during the dark hours, attacking the retreating North Koreans.[2] Gen. Douglas MacArthur's troops, flush with success, believed the war would be over by Christmas.

Wrecking his plans, Chinese hordes crossed the Yalu River and stopped MacArthur's drive in November 1950. The carriers provided both day and night air cover for the subsequent evacuation of the Marines from the Hungnam area ports. Before the crisis was over, the *Leyte, Princeton, Valley Forge* (after a short visit to the United States), and *Philippine Sea* furnished night combat detachments.

The first VC-35 detachment joined a VC-3 VFN detachment in the *Princeton*'s air group. Although there was no visible submarine threat to Task Force 77, the *Princeton*'s night attack team started operations by conducting antisubmarine patrols and courier flights in the Sea of Japan. By the middle of December the VAN detachment added their weight to VC-3's night air support effort. There was plenty of work for all the night flyers, helping the Marines retire from the Chosen Reservoir. The night teams kept dawn and dusk patrols armed to provide fire support over the Hungnam evacuation beach, staying on station from 19 December until the Navy evacuated all the Marines by the afternoon of 24 December. In daylight, their radar-equipped ADs sometimes guided the single-seat AD squadrons' strikes when cloud layers over the North Korean mountains made visual navigation to the targets impossible. Despite the night bombers' usefulness over land, the commander of Task Force 77

(CTF-77) had some of the night flyers back on antisubmarine patrol by 28 December.

The Chinese and North Koreans pushed the UN forces back to a line near the 38th parallel, where both sides dug in and began a war of attrition. As the new phase of the Korean War started in 1951, TF-77 commenced a series of air attacks interdicting elements of the enemy's transportation network, as well as hitting troop units and military installations in eastern North Korea. From the beginning of that year, the Navy kept an average of four carriers in the Korean theater until the end of the war, but the carriers conducted only 10 percent of their air operations at night. However, there were periods in which night operations were featured.

Operation Strangle was a planned air interdiction campaign conducted to stop Communist supplies from reaching the stalemated North Korean troops.[3] Navy squadrons began executing Operation Strangle in late May. The Communists quickly learned that their railroads and roads were now the prime targets for Allied air attacks. North Korean and Chinese trains and trucks began avoiding travel during the day, so it became the task of night flyers to locate and destroy vehicles moving on the North Korean rails and roads. The night crews also attacked the work parties sent at night to repair bridges, railroad tracks, and roads that were being destroyed during the day. The Communists added to the number of antiaircraft guns defending the supply routes, thereby increasing U.S. airplane and crew losses. The night teams, while flying fewer sorties than the day squadrons, took their portion of the losses.

During Operation Strangle, the night teams flew most of their sorties at night, taking off at dusk and during the early morning, predawn hours. The flight deck crew's workday lengthened from about fourteen hours to almost twenty hours to launch and land the night bombing missions. Only in the middle of the night after the dusk flights returned and before the predawn flights launched, did a carrier's Air Department men have any sleep or spare time.

・ ・ ・

The actions of the *Princeton*'s night teams were typical of all the VFN and VAN teams' activity during that year. During the spring of 1951, Cdr. Chick Harmer, commanding officer of VC-3, visited Cdr. Danny O'Neill's VFN team in the *Princeton* to get a close look at combat in his second war. Harmer went to the carriers because casualty reports started growing, usually from among the least-trained pilots. He thought he should get a feel for what was happening in this war. Harmer flew several night interdiction missions with the *Princeton*'s team, one of which he remembered well. On a bombing mission led by Lt. D. B. Shelton, Shelton got an ammunition truck at dusk right off the bat, but Harmer spent the evening dropping two 500-pound bombs and four 250-pounders on the blinking lights of the Chinese supply trucks without being rewarded by a resulting explosion or fire.[4]

The night flyers of VFN and VAN detachments usually launched two aircraft, a "section," on the night missions. Most section leaders separated their aircraft when the section arrived over land. The pair would operate singly, keeping in touch by radio, until one found a target or it was time to return to the ship. The APS-19 radar in the night Corsairs and Skyraiders was essentially useless for detecting small troop unit targets or even landmarks on the ground. The mountainous terrain added to the difficulty of using the airborne radars. The night pilots relied mostly on their night vision to find the enemy. Locomotives were sometimes easy to see even at night because of their smoke, but trucks used covered lights that the searchers could see only intermittently. However, at low altitude where they flew most interdiction missions on moonlit nights, pilots often detected vehicles, even those driving with covered lights along the roads.

If one of a section's pilots located a good target, the other would join the attack. Sometimes when the night air seemed too quiet and undisturbed to speak, pilots' radio conversations were whispers: "Psst, Owl Two, I've got a truck over here!" The second crew would fly toward the general target area guided by secondary fires from the first crew's attack or by parachute flares. The section leader would order the attack heading, pullout direction, altitude separation, and course for departing the target.

Other teams conducted all their night interdiction missions maintaining section integrity. One aircraft would fly at low altitude, searching along their assigned route, while the wingman stayed at about 5,000 feet, keeping the lead airplane's dim turtleback light in sight and navigating for the section. On particularly dark nights when it was unsafe to descend too low in the Korean mountains, both night attackers had to fly at 6,000 to 8,000 feet. If forced by terrain to higher altitudes, pilots had great difficulty seeing locomotive smoke or trucks.

The night pilots needed all their bombing skill to achieve the accuracy required to destroy small locomotives and trucks. Strafing with high-explosive incendiary (HEI) 20-millimeter shells was the preferred tactic for interdicting the enemy's transportation networks at night. After detecting a target, the crews hoped that a strafing run would set a fire and stop the vehicle so that bombing runs could follow. Crews preferred the 220-pound fragmentation bombs, usually with daisy-cutter fuzes, and 260-pound general-purpose bombs. They rarely used the heavier 500-pound bombs on interdiction missions. The teams used 1,000- and 2,000-pounders only on bridges or other large targets.

There were pros and cons about using flares to illuminate targets. Because night vision was essential, pilots were concerned about losing most of their night vision in the bright but diffuse light of the flares, but some pilots did rely on flares to assist them. All crews used flares as position indicators to establish the location of their section's aircraft if one found a target. When using flares for illumination, crews dropped the flares at an altitude calculated to allow the flares to burn out just as they hit the ground. The night flyers pulled out of their dives above the flares so that the enemy gunners would not see their airplanes silhouetted in the flare light.

The interdiction campaign against the North Koreans did not appear to affect their troops' battle strength. Because the fighting on the front was stable and relatively light, it was easy for the enemy to trickle needed ammunition and food through their railroad and road network to the front lines. Although the night flyers interdicting the transportation system did destroy many locomotives and trucks, the North Koreans replaced the lost rolling stock and trucks with oxen and manpower to move the needed supplies. The effort of the small number of night flyers helped little in keeping the enemy's front-line troops from accomplishing what they wanted.

Lt. (jg) R. C. Hessom of the *Essex*'s VAN Team 8 believed that the enemy had been slowed down a little because of night operations against their supply system, but that the night operations were too limited to affect the supply system sufficiently to make daylight transport necessary. However, some of the night fighters had a better opinion about their combat effectiveness. Rear Adm. Don Shelton reminisced about the success of the night work against the Communists, saying he heard the day guys at the O Club bar talking about kicking the Communists' butts, but in reality, the night flyers were the ones who really stopped the trains. Lt. Don Edge, now a retired captain, was proud of the skill of the night team pilots. He said that four VFN and four VAN could do more damage than an air group.[5]

. . .

The first AD-4Ns came to Korea as the Navy's night attack airplane in the summer of 1951. The crew positions were in the same place as the earlier night ADs: the pilot sat alone in his cockpit on top while the radar and ECM operators sat in the belly of the airplane. The winterized AD-4NL, which became the VAN detachments' standard airplane, followed before the end of the war, adding to the night attack flyers' effectiveness. Besides incorporating cold weather equipment, the AD-4NL introduced the APS-31B radar fixed to a wing station. It was more powerful and more reliable than the APS-19, with a scan of 135 degrees forward of the nose and a plan position indicator (PPI) presentation on the operator's scope. Figure 3 shows the difference in the distance presentation of the two types of scopes. There was no better air-to-ground radar in the fleet than the APS-31B until the 1960s, but it was no more effective than the APS-19 for detecting vehicles or troops, the critical targets of the Korean War.

The winterized F4U-5NL reached the VFN detachments off Korea in 1952 after being introduced to VC-4 two years earlier. Late in 1950, the squadron had sent a detachment to NAS Argentia, Newfoundland, to test the F4U-5NL's cold weather systems, which included rubber wing deicers, a fluid propeller deicing system, and a better cockpit heater. Icing conditions were sometimes so bad during flights through clouds that ice would build up between the cowling and the propeller on the non-winterized airplanes, scoring the metal propeller. The cold weather systems in the F4U-5NL and AD-4NL solved this problem, providing needed improvements the night combat pilots appreciated during the bitter Korean winters.

As a new year opened with the land war stalemated along the 38th parallel, a purely night air operation meant to block North Korean rail traffic, Operation Moonlight Sonata, began in the middle of January 1952. The operating carriers in the Sea of Japan launched ten night Corsair or Skyraider sections armed for interdiction missions at 0300 each night. During a three-hour mission, each two-plane section covered

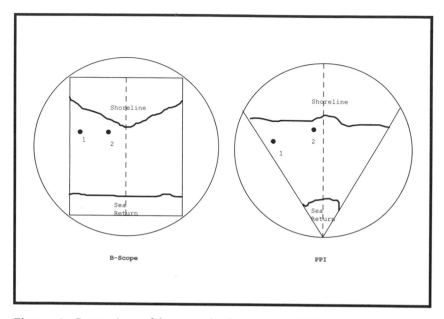

Figure 3. Comparison of the range displays on B- and PPI scopes.

about seventy-five miles of railroad track. The theory was that during the clear winter nights, the tracks and locomotives would be easy to see against the snowy terrain, but this night effort had marginal success. Moonlight Sonata continued through March, but the night flyers found only five locomotives and their trains, destroying two of the five locomotives.

After that operation, as this interdiction campaign wore on, the enemy truck drivers became smarter. They learned to extinguish all lights and stop when they heard an airplane. In turn, the night bombers started using more flares on a random basis. Road junctions and open valleys became likely spots to illuminate, with the hope of seeing trucks in the flares' light. The night raiders did locate and destroy a few vehicles using that random illumination tactic.

If there had been more night flyers, would the results have been better? Knowing what we know now about the overall success of interdiction of supplies in a low-intensity conflict, it is doubtful. At that time the North Koreans needed only small amounts of supplies at the front to sustain the war.

. . .

Flight conditions during the warm spring and hot summer weather were poorer than in the winter. Haze reduced visibility at night, blurring the mountains' shadows. Furthermore, the Allies had so many airplanes in the small North Korean airspace that midair collisions became a probability, not a possibility. Even at the speeds of the 1950s propeller-driven airplanes, if a carrier night aircrew lost their bearings, they could easily wander into the airspace reserved for Air Force operations in western North Korea.

When spring came in 1952, the night effort increased during a second night interdiction campaign. Operation Insomnia changed the night flyers' launch schedule and increased the number of sorties. Six sections launched from the carriers at midnight and recovered aboard after their night missions. Another six sections launched at midnight, landed at

The *Essex*'s flight deck with F4U-5Ns in snow during a Korean War winter.
U.S. Navy

61

K-16, an austere landing field near the battlefield, rearmed, and flew another interdiction mission before recovering on the carriers. Meanwhile, at 0200 a night's last six sections of VFN and VAN airplanes launched from the carriers to continue the night attacks.

During one night in May, a VAN section on the 0200 mission found eleven locomotives and claimed one destroyed and one damaged. Later in the flight, the night flyers stopped a sixteen-car train by strafing the engine and cutting the tracks. The flight leader relayed the location of the train to the task force commander. The train was in the Muchon area of North Korea, one of the best-defended areas of the country. At dawn, the carriers began a series of strikes on the train that lasted all day. In total, the day strikes destroyed nine engines and at least twenty-five boxcars. The North Korean antiaircraft fire downed one Corsair and damaged five others that day.

With their nightly effort during the two months of Operation Insomnia, the night detachments destroyed nine and damaged two of sixteen locomotives sighted. Considering both Operations Moonlight Sonata and Insomnia, the night flyers' success appears to be proportional to the number of sorties flown, not to the visibility conditions. However, the crews of the carriers that had their workday extended to twenty-four hours felt the strain of Insomnia at least as much as the enemy.

Throughout the summer of 1952, the joint command conducted an air campaign to break the stalemate in the war. Navy and Air Force squadrons flew mass air strikes hitting railroad, supply, and industrial centers bypassed earlier in the war. Complementing the large daylight strikes, the carriers' night teams kept flying interdiction missions to interrupt traffic on the rail lines and roads connecting the primary North Korean target areas. Two VAN teams, during a two-week period, claimed two trains and twenty-three trucks destroyed, plus sixty-eight trucks damaged, during night interdiction missions. Despite those successes, CTF-77 diluted the VAN teams' efforts by requiring them to furnish intermittent antisubmarine patrols.

North Korean air search and antiaircraft radars became more of a threat to the air groups in the fall of 1952. CTF-77 began scheduling the night ADs for passive ECM flights to locate the defenders' radars. During the first three weeks of September, the *Princeton*'s VAN team flew twenty ECM missions and only nine night interdiction missions. In early October, Chinese MiGs were first vectored against Navy aircraft, making electronic countermeasures even more popular. During that period, teams of day fighter or attack aircraft, guided by VC-35 ECM aircraft, flew several missions to locate and destroy North Korean GCI radar sites.

. . .

Because of the aircrewmen in the night attack airplanes, the deployed VAN teams were a little larger than the thirty-five-man VFN teams. The aircrewmen of VC-35 and VC-33 were unique—the only combat carrier crewmen in the Navy except for the crews in the heavy attack units. They worked in the cramped, dark, and cold fuselage of the AD night versions. To reduce light emitting from their aircraft over enemy territory, they blacked out the small windows on the belly doors, adding to their confinement. Only one small air scoop provided extra air for the radar, the ECM sets, and themselves. The pilot and crew members talked through an intercommunication system because the noise of the big R3350 engine made normal conversation impossible. The VAN team aircrewmen trusted their lives to the skill of their pilots just as did their predecessors in the World War II TBMs and other bombers.

A few years later I had an experienced radioman and radar operator in my AD-5N crew. He told a story about a carrier night landing after a Korean War combat mission. He was the radar operator in the belly of an AD-4N and had full confidence in his pilot. Preparing for this landing, he heard the normal sparse conversation between the LSO and his pilot and then was jarred by what he thought was the hard bump of a normal carrier landing. Then he felt water coming up over him in the dark fuselage. The airplane had caught and then disengaged an arresting wire in a skid and, rather gently, slid over the side. The airplane was still right side up in the water. Bond just unhooked his safety and parachute harness, opened his door in the darkness, and stepped out into the sea. A destroyer rescued Bond, the pilot, and the other crewman within an hour.

The night combat teams had been ignoring a category of targets that were important to the VT(N) squadrons in World War II: large industrial facilities. During the winter of 1953, Lt. Cdr. W. C. Griese, leader of VAN Detachment B in the *Valley Forge*, broke the monotony of the air interdiction campaign by executing a unique mission—the only documented carrier night strike on a North Korean industrial facility. One of his team's routine night interdiction routes passed within a half-mile of the Chosen Number One electric power plant, which had been repaired since the previous summer's strikes. The facility was near the northern end of the supply route structure that the team patrolled. Several night flights had gone by the power plant in the previous months without disturbing the antiaircraft gunners known to be in positions defending the plant. However, the task force had lost several airplanes to those same gunners during daylight raids on the plant. Griese convinced the strike planners that a night raid by his team would destroy all or part of the power plant with no losses.

One night in February, his flight of three AD-4Ns, each armed with one 1,000-pound general-purpose bomb, one semi-armor-piercing bomb, and six flares for illumination, launched at 0300 from the *Valley Forge*. Joining in a radar trail formation took a few minutes, then they

A flight of AD-4NLs over the Sea of Japan during the Korean War. *National Air and Space Museum*

64

flew the short overwater leg to landfall near Hungnam. Visual and radar navigation through the mountain valleys took them to the target about thirty minutes after takeoff. Griese found the power plant first on radar and then identified it visually in the dim moonlight as he flew by. Circling and climbing in the darkness, he dropped three flares. The second AD was the first to drop his 1,000-pounder a few seconds after the flares illuminated the plant. All three of the airplanes dropped the general-purpose bombs before repeating the bombing runs with the semi-armor-piercing weapons. When a flare burned out or hit the ground, the next airplane in the pattern dropped another flare. Although making repeated bombing runs, each airplane varied the dive heading to confuse the gunners. The night attack caught the antiaircraft gun crews napping, as expected. There was no ground fire at all until the last AD was in its final bombing run. Despite the success of this mission, there were no other carrier-based night air raids on industrial plants.

．　．　．

Although the air war in Korea was the focus of Naval Aviation for three years, there were significant events related to the improvement of the carrier's night and all-weather combat potential occurring outside of the theater of war. Simultaneously with the beginning of the Korean War in the summer of 1950, FAWTUPac and FAWTULant received their first jet airplanes. TV-2s—stable two-seaters derived from the Air Force's F-80s. The TV-2s were used to train fighter pilots in jet-powered airplanes' instrument flying and all-weather procedures and expose the FAWTU's night combat pilots to the differences between their propeller-driven fighter and attack airplanes and the jet fighters.

The Navy had started earlier in the year to get a night and all-weather fighter replacement for the F4U-5N to meet the future Soviet jet bomber threat. At the end of February 1950, Cdr. Harvey Lanham, now commander of Air Group Five, led a small group of F9F jet pilots landing at night for the first time on a carrier, the *Boxer*. Their experience showed that, despite the higher speed of that type of airplane, a jet night fighter for carriers was practical.

However, the Navy did not select the F9F as the next night fighter. In the spring of 1950, the Navy let a contract for delivery of fourteen F2H-2N Banshee night fighters to VC-4. By the end of 1951, VC-4 had received twelve of the F2H-2N aircraft, the only ones built. The Banshee night fighter was essentially the same F2H-2 aircraft that was flying in day fighter squadrons. It had the same straight-wing configuration, twin J34 engines with about 6,500 pounds of thrust, and four 20-millimeter cannon, but it had the F4U-5N's APS-19 radar in a slightly elongated nose forward of the cockpit. Just as in the F4U-5Ns, the radar image tilted on the scope whenever the fighter changed attitude because the radar antenna was not stabilized—but now the pilot had to work harder and quicker to interpret the radar picture because of the F2H's faster roll rate and the high closing speeds between the jet fighter and its enemy.

Selected as the leader for the first Banshee night fighter team, Lt. Cdr. R. H. Jennings picked Lt. (jg) H. A. Winter, Ens. W. J. Hepburn, and Ens. R. Smyth as VC-4's other jet night fighter pilots. The beginning of the war in Korea did not interrupt their training in jet night fighting. By the end of September 1950, the team was ready for night landing qualification and a short cruise to the Caribbean in the *Franklin D. Roosevelt* (CVB-42).

The *Roosevelt* was a straight-deck carrier, as were all the carriers of that era, although she was larger than *Essex*-class CVs. Performing carrier landings on a straight deck in a Banshee rated as the most hazardous of all the night flying tasks in naval aviation history. The jet's night or instrument approach started with a standard teardrop approach (by then practiced in both the Pacific and the Atlantic Fleets) from altitude. The Banshees then entered the landing pattern, flying by the starboard side of the ship on the landing heading. Turning downwind ahead of the ship and dropping gear and flaps, the Banshee pilot descended to 110 feet above the water, gauging his altitude visually, aided by the APN-1 radio altimeter that had not improved since F6F-5N days. He adjusted power and aircraft attitude for an airspeed of 105 knots (the day Banshee pilots flew the approach at 110 knots, but the night fighters wanted to be slower in case they did not catch an arresting wire).

At that point the pilot was almost ten football fields (a half-mile) abeam the ship. In low visibility it was hard to find the correct position abeam the ship's masthead light, the mark at which to start a turn into the ship's ramp. This was the critical time in the approach. The pilot, in pitch blackness or in weather conditions with no visible horizon, had to maintain his altitude, a constant speed, and a steady rate of turn. Meanwhile, he had to look sharply for the first sign of the LSO's signals, his only landing aid. The pilot usually did not see the LSO until after about

90 to 135 degrees of turn. At that point, the pilot began to feel that he might have a chance of getting aboard.

Lt. (jg) James A. Spargo was VC-4's jet LSO. He was a tall and lanky man who used the 1930s lighted wands rather than fluorescent paddles for signaling approaching airplanes. His outstretched arms, lengthened further by the wands, seemed to make his signals easier to see. He was even easier to see because of the phosphorescent stripes on his suit. For the approaching pilot on a dark and stormy night, he was a welcome sight. The first VC-4 Banshee team came back to Atlantic City after the Caribbean shakedown cruise with no accidents. However, the team had just twenty-eight carrier landings—only five at night—among the four team pilots.[6]

The first Navy jet night fighter became a showpiece for VC-4 and the Atlantic Fleet, but none were sent to the Korean War. VC-4's Banshee night fighter team deployed to the Mediterranean, where it was on station with the Sixth Fleet for the first five months of 1951.

The Banshee, in any version, had greater endurance and range than its contemporary cousins, the Phantom I, Fury, and Panther. Not only did the airplane carry more fuel but also the pilots had learned to shut down one engine, to "single up," while on fighter patrol stations or during flights to maximum range—an attribute that made it popular with carrier skippers. In contrast to the Pacific Fleet commanders, who at least tolerated carrier night and all-weather operations, Atlantic Fleet carrier captains and flag officers viewed night operations, even with the F2H-2N, as unnecessary. The Banshee may have impressed our NATO allies and the Russians, but VC-4's team averaged only one night landing per month per pilot—hardly enough practice for such a demanding task. Apparently, the commander of the Sixth Fleet, Vice Adm. "Cat" Brown, did not think the Russians would attack at night or in bad weather.

• • •

The F3D Skyknight was the second Navy jet night fighter. The first F3Ds went to VC-3 in 1951 during the middle of the Korean War. The F3D's two crewmen sat side-by-side in a relatively large cockpit. The airplane had the same two small J34 engines as the F2H-2N. It also had a large radar antenna in its nose, which caused high drag on its airframe. With its low power, despite being light at a gross weight of about 15,000 pounds, it could barely reach a maximum speed of 500 knots and was

VC-4 F4U-5N and F2H-2N in flight, 1953. *National Air and Space Museum*

not an agile airplane. In VC-3's evaluation report on the F3D, Commander Harmer stated that the Skyknight was a beautiful side-by-side two-seater and a wonderful cross-country airplane, but it was not a good carrier night fighter.[7] The primary problem was that the F3D took 30 minutes to get to 30,000 feet where the jet fighters and high-altitude bombers of the time were operating. Intercepts of future high-speed, land-based Soviet bombers attacking the fleet would be impossible unless the F3Ds were on CAP stations at all times. Moreover, VC-3 reported that it was difficult to land and operate on a carrier. The F3D had failed its first evaluation.

The Navy proved again that the Skyknight was slow for an interceptor while conducting a one-seat versus two-seat interceptor comparison test at VC-4 in the fall of 1951. The squadron performed the first operational test comparing the night fighter capability of a single-seat night fighter to the capability of a two-seat night fighter. The single-seat airplane was an

F2H-2N with a unique Hughes APG-36 radar. The radar was much more powerful than the APS-19 and was gyro-stabilized. The presentation no longer tilted as the airplane maneuvered, making it much easier for the pilot to interpret the radar picture of the sky before him. VC-4 pilots Merle Gorder, Bob Hunt, and Bob Smyth flew the special F2H-2N.[8] The two-seat airplane was an F3D flown by crews from VMF(AW)-323, Lt. Col. Tom Miller, commanding officer. For the test, the VC-4 and Marine aircrews flew under the control of a new GCI site at NAS Atlantic City. The simulated raids consisted of airplanes belonging to VC-4, AJs from the fleet, and JUs from the utility squadron at NAS Chincoteague, Virginia. All operations were at night or in adverse weather, or both.

The test proved more about the aircraft's radars than anything else. The larger radar dish in the F3D provided a longer detection range than that of the standard F2H-2N. Nevertheless, the Banshee finished first in the test by making its intercepts at a longer average distance from the hostile bomber's target because of its higher speed. The Hughes radar, however, was shown to be superior to either of the other radars. The stabilized antenna and picture on the pilot's display were the most important attributes of the Hughes radar. These specifications eventually became part of the night, all-weather fighter design requirements set by the Navy's Bureau of Aeronautics (BuAer). Although radar countermeasures were not used in the test, the pilots and aircrewmen who flew in the tests easily extrapolated that a second crewman, as in the F3D, might be absolutely necessary to cope with jamming or false signals in a countermeasures environment. This would prove to be the case in the next decades until the advent of computerized systems as used in the F-14 and F/A-18.[9]

In the face of criticism about the airplane as a carrier-based interceptor, VC-4 received the F3D in 1952 as a night fighter. VC-4 was changing to a jet squadron in that year. Three-quarters of the squadron's airplane inventory was propeller-driven in the middle of 1952; by December 1952, three-quarters of the airplane inventory was jet-powered. The squadron's main problem with the Skyknight was in maintaining its three radars. Because there were few qualified aircrewmen in VC-4, the men in the aviation electronics rating volunteered to become radar operators in the F3D. Their loss as full-time troubleshooters in the Maintenance Department further reduced the squadron's capability to keep the radars operating. Aircrewmen of all ratings, however, always did double duty, flying and working on the airplanes in their aircraft maintenance specialty.

VC-33 tried the Skyknight as a night attack airplane during the summer months of 1952. The tests failed to convince anyone that the rather slow, large, and underpowered F3D could do the night attack mission any better than the AD(N). In addition, the F3D radars were specifically tailored for air-to-air combat. The radar was ill-suited for locating targets on the ground.

. . .

One and a half years after the comparison test, the F3D deployed for combat with VC-4 Detachment 44 in the *Lake Champlain* (CV-39). Despite the general opinion that the F3D was not a good carrier airplane, the VC-4 pilots who took it to the Korean War liked the airplane even though it was ugly and underpowered. In particular, they liked its side-by-side crew configuration, its good search radar, and its good tail-warning radar, which they found valuable in combat. It was the Navy's only jet fighter to engage enemy aircraft at night during the Korean War. The Air Force night fighter of the time was the F-94 Starfighter, a tandem two-seater that was faster than the Skyknight but did not have as good an AI radar or any tail-warning radar. In the years working alongside the Starfighters, the Marine Corps and Navy F3D crews definitely showed their airplane's superiority as a combat night fighter.

Immediately after arriving in the Sea of Japan in June 1953, Lt. Jerry O'Rourke, Detachment 44's officer-in-charge, took the team ashore. The unit worked with the Marine Corps F3D squadron, VMF(N)-513, at K-6, another austere Marine airfield near Pyongtaek. Until the end of the war, the VFN detachment helped protect B-29 raids, engaging in several night dogfights with Chinese MiGs.

The MiG pilots and their ground controllers had developed tactics to destroy the B-29s bombing North Korea. Because the B-29s always flew at night, preferably on nights when there was no moon, the MiGs had learned the night game. The Chinese ground controllers would set the interceptors up for slashing attacks on the B-29 bomber stream. The Chinese fighters cutting into the B-29s hoped for a visual sighting of one of the big bombers so that they could get a good gun firing run.

To counter the night MiGs, the Skyknights were stationed on patrol stations at an appropriate distance from the Chinese border along the B-29s' route across North Korea from east to west. The Skyknights filled CAP stations every night, regardless of whether the B-29s flew or not. On nights when the big bombers did fly, the night fighters were ready well before the arrival of the B-29s. When an F3D, either while on a vector from Air Force controllers or flying under its own crew's control, detected a MiG closing on the B-29s, it turned toward the MiG. Usually that move was enough for the Chinese fighter controller to notice and order the MiG or MiGs to break north for the border.

There was another, much more lethal, variant to these feint-and-run matches between the F3Ds and the MiGs. The Chinese developed a tactic for destroying the U.S. night fighters while on their single airplane combat patrols. The Chinese trap started with a decoy MiG, the "rabbit," guided by the Chinese air controllers, approaching an F3D directly from the north. The Chinese expected that the rabbit would be quickly detected. When the Chinese noticed the patrolling F3D start an intercept and begin closing on the decoy, the rabbit would turn and gradually increase speed, never allowing the Skyknight to get within its cannon range. Simultaneously, the Chinese controller would vector in a flight of three other MiGs, usually in a spread formation, around the F3D.

The Navy flyers had a tactic of their own to counter the Chinese trap, however. As the MiGs attempted to close undetected for a shot at the Navy fighter's six o'clock, they often set off an audio warning signal from the tail-warning radar in the F3D crew's headsets. If a MiG was detected by an F3D's tail-warning radar, the Skyknight's crew would wait until the MiG closed to almost within its gun range. Suddenly the F3D's pilot would chop his throttles to idle power, removing the bright glow of the engine's fire in the airplane's tailpipe and most likely causing the MiG to lose sight of the Skyknight. Then the F3D's pilot would roll inverted and do a hard split-S maneuver, tumbling the attitude gyro if it was not caged, continuing up and around in an Immelman turn. At the completion of the maneuver, the pilot hoped to be within 20-millimeter cannon range on the tail of the MiG.

In the next few minutes of the engagement, the four MiGs, having lost the aid of their controller, and the F3D would weave through the night sky hoping to see their opponent and not run into each other. The MiGs also had to worry about shooting down one of their own. The F3D crew had some advantage over the MiGs in the beginning of the engagement because of their radars, and in the end game, an F3D could match a MiG in relatively slow-speed, turning dogfights. However, the Skyknight crews, just like the MiG pilots, had to spot fleeting shadows in the dark. Hardly ever were there firing opportunities on either side.[10]

Lt. (jg) R. Bick and an experienced radar operator, ATC L. C. Smith, began chasing a rabbit one night. They told their controller that they had fired on a target and had seen an explosion ahead of them. There were no more radio transmissions from the F3D. Presumably, the Chinese chasers surprised and destroyed Bick's airplane after he had destroyed the rabbit. Bick and Smith were credited with the probable kill of one MiG in the fight.

．．．

Another carrier-based night fighter unit moved ashore near the close of the Korean War. In July 1953, part of VC-3 Detachment D from the *Princeton* moved their F4U-5Ns to K-16 to shoot down obsolete North Korean biplanes that were too slow for the F3Ds to oppose. The Communists flew "Bed Check Charley" sorties almost every night, harassing Marine and Army units. With Marine Corps maintenance support and working under the control of Fifth Air Force, Lt. G. P. Bordelon, the officer-in-charge of the detachment, received a medal for destroying five slow-moving enemy airplanes during the twenty days of the war that Team D was ashore.

Another story from the last months of the Korean War concerns the fleet's night air combat capability. In late May about 0200 one night, the General Quarters alarm sounded in all ships of Task Force 77. The captains' big, vertical, circular air status boards showed blips detected by the air search radars of the *Boxer* and the other carriers in the force. Radar operators plotted unknown air targets on courses from North Korea toward the task force.[11]

In the almost pitch-black hangar decks of the carriers, aviation ordnancemen unloaded bombs already on airplanes for the next morning's strikes and quickly pitched the weapons over the side. The carriers with night flyers that were not already airborne launched the remaining night crews to intercept the raids. Other carriers launched some of their day fighters, who were ill-prepared for night combat. But then nothing happened. The blips on the scope disappeared as suddenly as they had appeared! Task Force 77 was probably the victim of radar ghosts caused by wet and cloudy weather.

The night fighters airborne that night had trained for such situations, but the day flyers that were launched had an unparalleled experience. Lt. (jg), later Rear Adm., Duff Arnold was on his first cruise in one of the day squadrons called upon for night intercepts. His assigned section leader that night was his skipper, an older Reserve Naval Aviator whom Arnold said was much smarter than he. The skipper claimed his radio did not work, so he did not launch. After the catapult shot, Arnold joined another section, flew around for a while following his air controller's vectors and landed back aboard after an hour or so of complete bewilderment and anxiety. That was Arnold's first night launch and landing ever![12] His experience was one that indicated how much better it would have been if all carrier aviators had been night qualified. At least the training would have given those young pilots some familiarity with the night world before facing it in a combat situation.

. . .

Beginning in late June 1953, the Allies increased the pressure on the enemy hoping to force an end to hostilities. The last twenty-eight days of the war saw the Navy's most intense air activity. For that month, with only four aircraft and five pilots each, the night teams typically flew four missions a night—quite an accomplishment. On 27 July, the Korean War ended after about a year of prolonged truce negotiations.

Naval Aviation flew about 130,000 combat sorties during that war, but Naval Aviators could not claim the destruction of a major naval power as they could after World War II. They could talk only about numbers of trains, trucks, and troops destroyed or killed. The four night squadrons sent twenty-four night fighter and twenty-two night attack teams to Korea. The night fighters destroyed six Korean airplanes at night, but only while operating with the Marines ashore in South Korea. The night interdiction missions flown by the VFN and VAN teams accounted for a high percentage of the locomotives, boxcars, and trucks destroyed during the Korean War. However, there was only one documented night strike on industrial facilities, airfields, or other major military facilities.

Although the Korean War was not as intense as World War II, it took its toll on the night flyers. In June 1952, the North Korean gunners shot down and killed Lt. R. J. Humphrey, a night ace in World War II. Humphrey, flying an F4U-5N, was team leader of VFN Detachment E in the *Princeton*. In total, the night flyers lost forty aircraft during the three years of the Korean War, an estimated six to seven losses per 1,000 sorties. This was a slightly higher loss rate than the overall loss rate in the war, about four aircraft losses per 1,000 sorties.

. . .

While the Korean War raged in the Pacific, the Atlantic Fleet's VC-4 detachments not only fought with three air groups in Task Force 77, but also furnished detachments in CVs, CVLs, and CVEs operating in the Atlantic or the Mediterranean Sea. The busy squadron got a tough leader when Capt. Josef M. Gardiner assumed command of VC-4 in 1951. He became known as a fierce advocate of night and all-weather combat capability for the aircraft carriers. As commanding officer of VC-4, Gardiner supported his team leaders, trying to get more opportunities for them to exploit their night-fighting potential. He felt strongly that the Navy was the loser because of the neglect of night operations that was so prevalent in the Second and Sixth Fleets. Gardiner kept up a stream of advice for his seniors, stressing the need for carrier captains and flag officers to keep the night teams proficient. At the time, his words did not help correct the situation, but his opinions made an impression on the next generation of fleet commanders.

In the early 1950s, presenting a nuclear weapons threat to the Communists began to shape the Atlantic Fleet's aircraft carrier activities. Acceptance of Captain Gardiner's arguments for an effective night and all-weather combat capability was certainly needed as the Navy continued efforts to put an effective nuclear weapons potential in carriers. August 1950 was an important month for the first all-weather, nuclear weapons squadron. All VC-5 crews completed carrier qualifications in the AJ-1 while flying off the *Coral Sea*. Having trained their own pilots and aircrewmen, the squadron was ready for sea duty, adding to the United States's nuclear strike capability.

The AJ squadrons' aircrews performed some of the most dangerous flying ever attempted while flying from carriers at night. Their operations from the straight decks of the *Essex*- and *Midway*-class carriers used paddles passes to get aboard and matched in difficulty the task of the Banshee pilot, but for different reasons. Unfortunately, the Savage was only

Capt. Josef Gardiner and Lt. G. G. "Jerry" O'Rourke with VC-4's Detachment 44
In front of an F3D. *U.S. Navy, courtesy Capt. G. G. O'Rourke*

barely suitable for carrier operations. Although the airplane was relatively slow in the landing configuration, it was the biggest airplane ever
to attempt carrier night landings up to that time. Frequent night flights
were probably more essential to maintain the proficiency of AJ crews
than for the pilots of the smaller airplanes because of the small margin
for error pilots had while landing that large airplane on small, straight
carrier decks. Moreover, the crews selected to fly the AJs were from shore-
based patrol plane squadrons. Patrol plane aviators did have large
amounts of night and adverse weather flight time, but no carrier experience. They showed their lack of carrier experience trying to land their
new airplane. It was common for everyone in a carrier's air group,
mostly day fighter and attack pilots, to spend time in the catwalks on the
carrier's island superstructure watching the Savages come aboard at
night. After observing a series of the AJs' night landing operations, Vice

Admiral Brown, while commanding Sixth Fleet, strongly suggested that the heavy attack squadrons must receive experienced carrier pilots in the future.[13]

Despite their difficulties, the heavy attack squadrons quickly, but gingerly, began to use their new capability. VC-5 reported that in 1950 the squadron flew over 2,000 hours, of which about 180 hours were flown at night. The next year the number of hours flown in the AJ had grown to 3,000, with 200 at night.

VC-5 became the first squadron in Heavy Attack Wing 1 (HATWING 1), established on the East Coast in February 1951. VC-6, established in January 1950, and VC-7, which had been established at NAS Moffett Field in 1949 but moved to the East Coast, added two heavy attack, nuclear weapons squadrons to HATWING 1. During 1951 and the next year, AJ detachments deployed in the *Coral Sea, Franklin D. Roosevelt,* and *Midway*—the three largest aircraft carriers in the fleet—helping to meet the Navy's requirement for two carriers in the NATO area. To meet a crisis situation in the Mediterranean during March 1951, VC-5 deployed AJs to Port Lyautey, French Morocco, for further transfer to CVs if necessary. The carriers were ready at a moment's notice to launch the AJs armed with nuclear weapons against targets in the Soviet Union or the Warsaw Pact nations as assigned by NATO commanders.

Developing tactics and weapons-delivery techniques as they became more proficient in the AJ, the heavy attack crews perfected a high-altitude, level-flight bombing procedure using the airplane's air-to-ground radar and bomb director system. In one strike exercise in November 1951, a VC-5 detachment performed four simulated weapon drops at night, together with five daylight weapon drops. The exercise proved the capability of the squadron and the AJ to perform its mission day and night.

Radar scope photography, necessary for improving the navigation and bombing accuracy of the night and all-weather attack crews, also gained in importance. Many of the AJs' missions involved taking radar scope pictures of possible target areas in Europe. The three-man crews flew their long-range missions in any weather, but they suffered from lack of attention by the operating task force and fleet commanders, just as the night fighter units did.[14]

VC-6 moved to NAS North Island, becoming the first heavy attack squadron in the Pacific Fleet in January 1952. VC-6 pilots completed their carrier landing qualifications on the West Coast in August 1952. From that time, the squadron furnished detachments to carriers deploying to the Western Pacific. During October 1953, a VC-6 crew dropped an Operational Suitability Test weapon (a full-scale nuclear bomb, but without an nuclear warhead) on a radar-reflecting target. The crew released the weapon in level flight from 26,000 feet using only their all-weather radar and bomb director system, achieving a hit that was within acceptable limits for the test.[15]

Nuclear weapons had become smaller in weight and volume by 1950, allowing small carrier airplanes such as the AD and the Banshee to carry the new class of nuclear bombs. By the summer of 1952, VC-3 and VC-4 had a version of the Banshee—the F2H-2B—capable of delivering nuclear weapons, and VC-35 had AD-4Bs for that purpose. Although VC-33's VAN teams sometimes flew the old TBM-3E because of a shortage of the better AD-4Ns, during the winter of 1952 VC-33 also added the nuclear weapons delivery mission to its responsibilities. After 1952, the squadron deployed at least two AD-4Bs with their teams in the Sixth Fleet. These squadrons were selected for the mission because of the need to conduct those missions at night and in any weather condition. However, the final delivery of the weapon with these airplanes required visual conditions in the target area, leaving the AJ as the only true night and all-weather nuclear weapon attack airplane in the Navy.

During October 1952, four F2H-2Bs in VC-3 Detachment GB joined the VC-35 detachment already based at NAS Atsugi, Japan, for nuclear weapons missions, if necessary. The theory was that the VC detachment pilots and aircrews could be switched between the shore-based nuclear weapons teams and the night fighter and attack teams assigned to the carriers fighting the Korean War. In so doing, the carrier night operations as well as nuclear weapons delivery proficiency of all the VC squadrons' combat pilots would be maintained. The pilot switch was to occur when the combat crews needed rest. In hindsight, as might be expected, the stress of battle tended to make the men sent to Atsugi neglect their nuclear weapons training. After the Korean War ended in 1953, the VC-35 and VC-3 nuclear weapons detachments at NAS Atsugi disbanded.[16]

Lt. (jg) A. A. Schaufelberger gained his first night experience as a

VC-33 night attack pilot during this period. He reported to the squadron in 1951 directly from flight school, without going to FAWTULant, and later went on to become one of the first of VC-33's nuclear weapons pilots. In the spring of 1952, Schaufelberger deployed as a member of VC-33's team in the *Wasp* (CV-18) on a cruise to join the Sixth Fleet. One night en route to the Mediterranean, several hundred miles south of Iceland, the task group's destroyers were reorienting their antisubmarine screen forward of the carrier as the CV changed course to recover some of her night flyers. During the maneuvering, the *Wasp* and the destroyer-minesweeper *Hobson* (DMS-26) collided. The collision sank the *Hobson* and severely damaged the *Wasp*'s bow, limiting the carrier to a top speed of only a few knots.

Schaufelberger and two other VC-33 pilots were airborne when the accident happened. Because they did not have enough fuel to reach Iceland, the closest point of land, the pilots had the choice of ditching their airplanes at night in the cold Atlantic waters or attempting a landing. The VC-33 pilots chose to land aboard the *Wasp* with only a few knots of wind over the deck. Allowing for safety margins in the arresting gear wires and the airplane's structure, the AD required about 25 knots of wind over the deck for landing. With the *Wasp* barely making steerageway, the VC-33 pilots landed that night with only 10 knots of wind over the deck as the arresting gear crew tensioned the wires to the highest pressure. All of the pilots landed safely, stopping before touching the barriers on the carrier's straight deck. That episode shows that the VC squadron pilots had exceptional talent in controlling the airspeed, attitude, and altitude of their airplanes—a skill absolutely necessary for night air operations on straight decks.[17]

• • •

Captain Gardiner continued his efforts to advance his theories throughout his command tour at VC-4. In the spring of 1953, he and a group of forty officers from his squadron, as well as aircrews from VC-33 and VC-12, joined in the *Roosevelt* to demonstrate the offensive and defensive capability of the night combat units. The small task unit conducted night fighter and attack missions during a period of thunderstorm conditions off the Virginia Capes. The aviation flag officers did not view the demonstration with enthusiasm because, among other reasons, VC-33 lost one airplane and crew during the short, four-day exercise. The admirals criticized the night, all-weather exercise as being ill-conceived and the basic cause of the loss. On the other side of the argument, Captain Gardiner explained again that the Navy had to have a twenty-four-hour combat capability. He placed the blame for the accident on (1) his superiors' lack of support for proper training and (2) lack of better carrier and airplane systems for night and all-weather tasks.[18]

Despite the judgments of some flag officers, it was becoming increasingly apparent that the Navy must develop a true twenty-four-hour air combat capability in her carriers because of the Navy's increasing responsibilities in the Cold War. During a large NATO exercise in the Atlantic in 1953, an incident occurred that vividly illustrated the need for a better night and all-weather capability in carrier aviation. About one hundred miles south of Greenland in the middle of winter, the *Bennington* (CV-20) and her support task group readied to launch all her ADs in a mock attack against a simulated Soviet fleet steaming out of the North Sea. The sky was gray with thick clouds and fog was forecast.

The first plane off the *Bennington* was the VC-12 detachment officer-in-charge, Lt. Cdr. Jim Wiggins, in an AEW-configured AD. Wiggins stayed under the overcast, where no more than twenty miles from the ship he ran into a solid and thick fog bank that he duly reported back to the carrier. The *Bennington* continued the launch, which luckily included nothing but ADs that had fuel to stay aloft for many hours. The ship was still launching when she ran into zero-zero visibility and ceiling conditions in the fog bank. The ADs that were airborne started an orbit in a clear area downwind of the pea soup fog. As the fog moved downwind, the *Bennington* could not outrun the mist and still have room to turn into the wind and recover aircraft. No breaks appeared in the fog while the aviators on top of the fog bank and in the *Bennington* steaming in the fog were searching for open areas.

Nightfall approached and the ADs' fuel ran low. Because all the ADs were out of range of a landing field in Iceland or Greenland, a submarine assigned to the exercise prepared to rescue pilots when their fuel exhausted. The airborne pilots, led by Cdr. Ben Preston, commanding officer of VA-75, took a vector toward the submarine, but en route Preston noticed a hole in the fog below. Simultaneously, the carrier steamed into the same area, which had low a ceiling but good visibility. It did not take much urging for the AD pilots to drop through the hole and set up a landing pattern. The *Bennington* turned into the wind and began recovering aircraft. The hole through the fog closed in a few minutes, but the area of good visibility underneath remained. As darkness fell over the ocean, Preston had to return on top of the fog several times to gather inexperienced pilots on his wing to guide them through the clouds. Miraculously, no airplanes ditched in the North Atlantic.[19]

. . .

VC-4's first trial with the prototype F2H-2Ns showed the promise of a night fighter with higher performance and better flying qualities for carrier work than that airplane or the F3D. The F2H-3 was the third jet night fighter to fly from carriers. It was essentially a basic F2H, but with a larger fuselage. At a gross weight of 14,000 pounds, but still with a straight wing, it had a maximum speed of less than 550 knots. The airplane's day fighter contemporaries, both Navy and Air Force, were faster. The dual-mission night fighter/nuclear weapons capable Banshee had more fuel and weight in its fuselage extension than the earlier F2Hs. It carried four 20-millimeter cannon and had ten wing hard points for nuclear weapons, conventional bombs, or rockets. Unfortunately, the new Westinghouse APQ-41 AI radar was unreliable and still used an inefficient, unstabilized radar scan and scope presentation. It did not solve the single-seat night fighter's major problem: maintaining radar contact with an enemy while maneuvering.

The next Banshee version followed the F2H-3 to VC-4 in the spring of 1953. This airplane, the F2H-4, though still a Banshee, was a real improvement as a night fighter because of the Hughes APQ-37 radar (E-10 fire-control system). The APQ-37's radar power was greater than that of earlier radars, providing longer detection ranges. The APQ-37 was the first Navy operational aircraft radar with a space-stabilized antenna. The Navy backfitted about half of the F2H-3s with the new radar because it was such a great improvement over the old system. The pilot's radar display, with a larger six-inch scope plus its stabilized horizon line, made interpretation of the radar scene before him much easier. There was also a steering circle and steering dot that made it simpler to see the target's relative bearing and altitude from the Banshee. The appearance of the steering dot indicated that the radar had "locked on," that is, the radar automatically kept its beam on the target; the pilot did not have to control its beam to track the target. With its simple lead-computing radar system, a successful intercept was much more certain than in the earlier night fighters. Another feature of the F2H-4, also backfitted into the F2H-3, was the first "tumble proof" attitude indicator, placed in the center of the instrument panel. This instrument was supposed to keep an attitude reference for the pilot while completing such overhead maneuvers as a loop or half Cuban-eight during darkness or in clouds.

Neither version of the "Big Banshee" saw duty during the Korean War. Although coming to the squadron in late 1952, VC-4 did not deploy the F2H-3 until May 1953, and then only to the Mediterranean theater. The F2H-3 first deployed with a Pacific Fleet VC-3 detachment in July 1953, arriving in Korean waters about a month after the truce.

Lt. Cdr. Bob Lyon led the first VC-4 F2H-4 team. Lt. R. C. Stewart, Lts. (jg) R. D. Harris, Jr., J. K. Setliff, and Otis T. Inge, and a maintenance crew of about sixty men filled out his team. The team took four F2H-4s to the Mediterranean in the *Bennington*, leaving Norfolk, Virginia, 30 August 1953. Lyon described one mission during that Mediterranean cruise with Sixth Fleet:

> Flying in the Med atmosphere was usually in a "milk bowl"—that is, so hazy that one sees no horizon, no water, no sky. For reasons unknown, three of us were launched one night as a division. Following climb-out and completion of the assigned mission, *Bennington* advised us to stay aloft [and conserve fuel] until a foul deck was cleared. Later we were ordered down for landing. Having descended, we were ordered to orbit in vicinity of the ship because the foul deck had not been cleared as anticipated. Weather was deteriorating, with a low ceiling developing and visibility reducing. A flight of ADs were milling about, having also been directed to orbit in vicinity of the ship. At low altitude our division skimmed beneath cloud cover as fuel reserves dwindled. The AD flight passed dangerously close under us in a near-miss. When we finally got back in the ready room, I asked Otis Inge, "How much fuel did you land with?" Otis [replied,] "The aircraft ran out of fuel before I could park it."[20]

Lyon recalls another personal incident in a Big Banshee during that cruise. The story emphasizes how important every detail of flying was—and still is—to the safe and successful completion of each flight. Although experienced in high-performance, propeller-driven fighters, Lyon was one of the first to fly the jet night fighters. Contemporary VC-3 accounts tell of short and skimpy checkout procedures in the new jets.

VC-4's checkouts were not any better. Although Lyon's team had flown the F2H-3 for almost all of their predeployment training cycle, the added speed and mission altitude differences between the Corsair and the Banshee required significant changes in the pilot's flying habits. It is understandable how the following incident could happen to a good Naval Aviator.

One night after letting down through the clouds, I missed seeing the ship's masthead light on entering the landing pattern. In the descending, standard-rate left turn downwind, I lowered the wheels and flaps. At 150 feet of altitude on the downwind leg, when the radio compass, "Bird Dog," pointed exactly ninety degrees to the left, I glanced through the left side of the canopy and saw the masthead light. All was well as I commenced the final standard-rate left turn. The radio altimeter checked out with the adiabatic pressure altimeter at 90 feet. Turning up the groove, I stole a glance from the instruments to see if the deck was in sight. The only visible things were in the cockpit! In the left turn up to the groove, suddenly phosphorescent water, the ship's wake, was apparent in the darkness below. Quickly lining up on the wake, the view through the left side of the canopy changed from phosphorescence to a cut signal from the lit-up LSO, who miraculously materialized 35 feet down to the left [at the ramp]. Hauling off throttles was followed by the negative g's of an arrested landing.

As the plane came to a stop and drifted back a few feet, the flight deck bull horn blasted, "Turn off your lights!" Lighted wands directed my taxiing forward over the lowered barricade. Wing folding, parking, and engine shutdown all came in rapid succession. Being in the chocks again seemed premature, as if I had not caught up to where I was. An astonished plane captain clambered up to the cockpit to inquire about an inch of frost covering the windscreen. Realizing then the reason for not seeing through the windscreen, I admitted, "Forgot the defrost." After that I remembered to defrost when descending through clouds.

Another incident reflecting the trials of a night flyer happened to a VC-4 pilot on another cruise in the *Wasp*. At night, there is always the possibility of losing instrument lights. Night flyers learned early to have a working flashlight handy in the cockpit. After a night navigation flight, Lt. G. H. Webster descended along the starboard side of the *Wasp*, turned downwind, and dropped landing gear and flaps. As he moved the throttles forward to stop his descent, he lost power in one engine. The Banshee could not maintain its altitude with its weight and only one engine. Webster figured that the circuit breaker controlling the vital fuel pump for the failed engine had opened, stopping fuel flow to that engine. He stripped off his oxygen mask and put his flashlight in his mouth. Needing light to see the circuit breaker panel down by his feet and two hands for conducting the engine relight procedure, he had thought fast enough to use his mouth as a flashlight receptacle. Resetting the circuit breaker allowed him to relight the engines and continue the flight to a safe landing.

Chapter 5

CARRIER NIGHT AND ALL-WEATHER OPERATIONS MATURE

By the end of the Korean War the four night fighter and night attack squadrons were huge. The number of pilots, aircrewmen, and maintenance personnel in each squadron numbered between five hundred and one thousand men. The few heavy attack units added to the night flyers. With eight to twelve night-equipped airplanes in each deployed air group, the night combat squadrons had established their niche. However, the number of night and all-weather aviators remained small in comparison to the total number of carrier aviators. Flying at night and in any weather condition from an aircraft carrier was still considered an elite specialty in Naval Aviation. That situation would soon change as Naval Aviation took advantage of new ideas and better ship and airplane systems.

The early pioneers of carrier night air combat were rapidly passing from the scene and would not see the fruition of their efforts to establish a universal night fighting force. Bill Taylor finished his career during World War II. Gus Widhelm taught for a while at Barbers Point, then retired. Chick Harmer and Bill Henry left their final marks as night fighters on VC-3 during the Korean War. Other World War II and Korean War–era leaders, as well as the night aces, contributed their efforts but moved into other fields. Turner Caldwell passed on his experience at NACTUPac before moving on to other achievements in Naval Aviation and to the rank of vice admiral. A few other night air combat pioneers had been promoted to flag rank by 1950. Rear Admiral Ewen established the carrier's presence in the Sea of Japan and started night air combat operations when the Korean War began, but left the Navy soon after the war. Vice Admiral Gardner led Naval Aviation in the CNO's office for a period during the Korean War.

In support of the fleet's aviators, from World War II to the middle 1950s a group of technically inclined aviation admirals had directed the development of aircraft carriers, carrier systems, aircraft, and aircraft systems that promised to improve aircraft carrier operations. The first of the significant carrier modifications went to sea in 1953 when the Navy brought into the fleet a system that significantly improved the safety of carrier night and all-weather operations.

In an experiment with carrier flight deck design, British naval architects had offset the deck's landing area at an angle from the fore-and-aft axis of the ship. That made it possible for airplanes that missed an arresting wire to touch down and immediately take off again, removing the danger of a crash into barriers protecting parked airplanes on a straight deck. The U.S. Navy wisely accepted the angled deck as a modification for its carriers. The *Antietam* (CV-36) entered an Atlantic Fleet shipyard in late 1952 and came out in January 1953 as the first angled deck carrier in the U.S. Navy.

Electronics systems and weapons also improved by 1953. Capt. Charles T. Booth II, who had served as commanding officer of VC-4 before Captain Gardiner, took charge of the Electronics Test Directorate of the Naval Air Test Center in 1951. His leadership in the next two years helped improve the night and all-weather potential of carrier aircraft. The Navy began equipping carrier airplanes with TACAN and multi-channel UHF radio sets. New, greatly improved air intercept radars were developed. Designers also made advances in air-to-ground radars, radar altimeters, and attitude indicators for night and all-weather fighter and attack aircraft. An event that changed the night fighter world occurred in February 1953, when the Navy conducted live firing tests of the Sparrow I, the first radar-guided air-to-air missile. With an effective radar missile, night fighters could now finish night engagements beyond gun ranges that had previously almost required a midair collision to get a good shot at the enemy. In addition, with radar vectoring from the ship's air controllers and with a good air intercept radar, night fighters could complete engagements in clouds and some rainy weather conditions.

To increase the night flyers' effectiveness using the improved night and all-weather systems, there was more emphasis on instrument flight training. By the summer of 1953, all student aviators passed through the All-Weather Flight School to earn the Navy's standard instrument rating. However, nuggets assigned directly to night air combat units still had no tactical night flying training. A core of individuals who worked through two wars to build the carriers' night and all-weather capability stayed on as teachers for the next generation of Naval Aviators. Future vice admirals Bill Martin, today praised as the father of carrier night aviation, and Pete Aurand, as well as some of the junior officers who pioneered instrument flight in the early jet fighters or flew in the VC squadrons during the Korean War, continued to improve the effectiveness of carrier night operations during the late 1950s. Undoubtedly, by the end of the Korean War, Naval Aviation commanders could see ahead toward a fleet of larger, safer carriers more capable of supporting the need for seaborne night and all-weather air combat operations in opposition to Soviet aggression.

. . .

With the war over, the Navy's primary mission focused on showing the naval nuclear weapon potential to the U.S.S.R. and its ally, Communist China. That mission shaped all the aircraft carriers' and the night combat squadrons' operations until the beginning of the Vietnam War. The fleet commanders expected the heavy attack squadrons to be principal contributors to U.S. nuclear strike plans prepared to deter major acts of Soviet aggression.

Despite the problems with its airplane, Heavy Attack Wing 1 continued nuclear attack training operations at NAS Sanford, Florida, and aboard Atlantic Fleet carriers throughout the middle 1950s. Heavy attack pilots flew paddles passes on the straight decks from 1950 to 1956—a long, difficult time for AJ crews. Crowds still turned out during the 1955 *Oriskany* (CV-34) cruise to watch the AJ's night landings that continued to challenge the heavy attack pilots.[1] On the West Coast, the Navy established squadrons VAH-6 and VAH-2 in 1955 to provide Pacific Fleet *Essex*-class carriers with AJ bomber and tanker detachments. Heavy Attack Wing 2 formed with those squadrons at NAS Whidbey Island, Washington, in July 1956.

The night composite squadrons also continued to contribute to the nuclear strike plans. VC-4 provided the Atlantic Fleet carrier air groups with VFN teams, flying either F2H-4s or F4U-5Ns, without interruption after the Korean War ended. A typical VC-4 night fighter and nuclear weapons detachment, led by Lt. Tony Feher, worked in the Sixth Fleet from October 1954 to March 1955. Feher's detachment flew Big Banshees off the *Lake Champlain*. The Navy had modified some AJs to be aerial tankers to provide fuel for the carriers' jets. The *Lake Champlain*'s air group had one of the first tanker- and bomber-equipped VC-5 detachments. Having extra fuel available in the air eased the fuel reserve problem of the air group's jets. The F2H-2B and the Big Banshee were the first fighters equipped with an aerial refueling system. Flying the Banshee into the drogue at night was a new technique that required practice, but the result—getting more fuel—was always worth the trouble.

The *Lake Champlain*, like all the carriers except the *Antietam*, still had a straight deck blocked by barriers. Furthermore, carrier night air control and landing procedures were essentially the same as before the Korean War. With those limitations, the carrier captains and the carrier division staffs still did not appreciate the value of night operations. The crew's nightly movie, shown on the hangar deck, took precedence over night flying. The five pilots in Feher's detachment averaged only four night landings in a six-month deployment,[2] decreasing their effectiveness and increasing the possibility of accident when the night fighters did fly. The commanders' attitude set up some deadly situations at times. On their way back across the Atlantic after the Sixth Fleet deployment, the *Lake Champlain*'s captain decided to hold flight operations one night after a long layoff for the air group. Two VC-4 pilots sat in their Banshees that night, hardly able to see the catapult officer because of the rain in a thunderstorm. At the last minute before launch, the captain canceled night flying.[3]

Pilot transition from props to jets in the 1950s was informal and unstructured. Airplane flight manuals contained only a few pages. Lt., later Capt., Don Edge had only eighty-five jet hours before deploying with a Big Banshee VFN team. A few months after Cdr. J. D. Ramage took over as VC-3's commanding officer in 1954, he set up a jet transition syllabus for his pilots. Ramage's practice of using a supervised training course for pilots changing to higher-performance or different aircraft started a trend toward better safety procedures in carrier aviation.

The *Bennington*'s deployment in 1955/56 included VC-4's Detachment 30, led by Lt. Emmett Cooke. The ship had received an angled deck in

The *Lake Champlain*'s VC-4 night fighter team in formation over the Mediterranean Sea. *National Air and Space Museum*

1954, making night landings easier for the Big Banshee pilots. The *Bennington* steamed into the Western Pacific, where the team participated in two nuclear weapons exercises near Okinawa and Iwo Jima. The VC-4 detachment's deployment report states that the team flew 142 hours on weapons delivery practice flights, including thirty-eight hours from shore bases at NAS Atsugi, Japan, and NAS Sangley Point, the Philippines. Detachment 30 was the last VC-4 Banshee team deployed in carriers.

. . .

The Navy took a big step toward removing the "elite" label from the carriers' night operations in 1954 when the Navy designated about fifteen day fighter squadrons—one per carrier—as night and all-weather fighter squadrons. All but two of the new night squadrons received dual-mission Banshees like those flown by VC-3 and VC-4. For about two years, those new squadrons shared the missions flown by the VC-3 and VC-4 Banshee detachments. The nuclear weapons delivery mission was foremost in their tasking, followed by night fighter tasks. The squadrons also trained for night dive bombing using conventional bombs. The Big

Banshee squadrons enjoyed the heritage of the World War II night Hellcat squadrons and the night fighter squadrons flying for a short time in the late 1940s. The wholesale reduction of Naval Aviation foiled the postwar plan for night fighters, but now the fleets were ready to expand the carriers' night air combat potential.

Alongside the new F2H squadrons, the fleet again tried the combat-proven F3D for carrier duty when, in 1954, the Navy assigned a full complement of F3Ds to VF-14, joining the Big Banshee night fighter units. The squadron's mission combined night intercepts and night attack, but it did not have a nuclear weapons mission. VF-14's pilots and radar operators flew the Skyknight until 1957, operating on Atlantic Fleet carriers, but not making any long aircraft carrier deployments.[4]

VF-82 was another squadron flying unique airplanes during the fleet's transition to Big Banshee squadrons. It received six F2H-2Ns and six F2H-2Bs from VC-4. The entire squadron, previously a day fighter squadron, deployed to Key West for a month to train under the supervision of FAWTULant's night fighter instructors before starting night intercept work in the squadron. VF-82 shared the 1954/55 Mediterranean deployment in the *Lake Champlain* with Feher's VC-4 detachment. VF-82's night fighter pilots experienced only about forty carrier landings during that cruise and no one in the squadron performed more than three night landings. The squadron's leaders noted that the quality of its aviators' ability for night work varied significantly. The training officers worried about the design of the night and foul weather teardrop approach and entry into the paddles landing pattern because the less-experienced pilots had much difficulty performing that maneuver safely. The squadron tried alternative procedures, including the downwind break used by some of the VC teams, but nothing different seemed better than the old procedures.[5]

VF-13 and VF-21 were two of the first Atlantic Fleet squadrons to fly the dual-mission Banshee. Those squadrons deployed in the *Franklin D. Roosevelt* and *Coral Sea*, still straight-deck carriers, in 1955. VF-71 was another Atlantic Fleet squadron flying the F2H-4. Under the command of Cdr. Jim Donnelson, the night fighter and nuclear weapons squadron joined the *Hornet*'s cruise in 1955. VF-71 flew across the country to join the ship at NAS Alameda, California, before departing for the Western Pacific. In the Pacific Fleet, VF-114, VF-124, and VF-193 were early squadrons with the dual-mission Banshees deploying in the *Essex, Lexington,* and *Oriskany* in 1955 and 1956.

Lt. (jg) J. R. C. Mitchell reported to VF-193 immediately after completing flight training during the summer of 1954 as one of the new aviators sent directly to night air combat squadrons. Mitchell became one of FAWTUPac's last students in its advanced instrument and night fighter training syllabus, which had not changed significantly since the Korean War. The role of the FAWTUs in each fleet had begun phasing out in 1953. The FAWTUs' night combat courses ceased completely in 1955. From that time on, the fleet squadrons had to teach all the advanced night and instrument flight techniques, plus night air combat tactics.

Cdr. Mickey Weisner moved from VC-3 to become commanding officer of VF-193. He decided that the new pilots reporting to his squadron would have enough to do learning carrier night operations and only one of the two missions, night fighting or nuclear weapons delivery. Consequently, the new aviators spent their first year practicing only a single mission, but the senior pilots in VF-193 performed both tasks. After a short squadron training period, the squadron deployed in early 1955 to the Western Pacific in the *Oriskany.* The air group had two other night combat units, a heavy attack squadron flying AJs and a VC-35 detachment for night attack and nuclear weapons delivery.

There were two incidents on that cruise that show the hazards of carrier night aviation in the mid-1950s, when the majority of the carriers still had straight decks and barriers. The first featured a VF-193 pilot who was on the last flight of the night. It was common practice in night squadrons to turn out the ready room's red lights and begin a movie soon after the last flight finished its briefing and left to man airplanes. Returning to the ship after a practice intercept flight, the night flyer missed the wires on his landing attempt and flew into the barriers. The crash threw the airplane down on the flight deck at the spot right over VF-193's ready room. Most of the pilots and maintenance chiefs were watching the movie when the crash occurred with a Wham! overhead. Dust, debris, helmets, flashlights, and anything else loose or not secured flew around the room. Now in the dark, the pilot's squadron mates were as startled as he was by the crash. After the confusion abated some hours later, Commander Weisner issued an order that no movies would begin until after the last night fighter was aboard.

The second incident happened to Lieutenant Mitchell near the end of this WestPac deployment. An excellent pilot, Mitchell was a victim of the toughest flying environment that ever existed, bringing a jet fighter aboard a straight-deck carrier at night with only the help of an LSO wearing a faintly illuminated suit. One exceptionally dark night, after a fair pass around into the groove, he began losing altitude, coming in critically low. Frantic "low" signals from the LSO did not get a response. Mitchell crashed his Banshee, starboard wing down but straight ahead, below the flight deck into the fantail "potato locker" of the *Oriskany*. Mitchell's starboard wing hit one of the fantail's 40-millimeter gun tubs and separated from the Banshee. The forward section of the airplane carried the cockpit and Mitchell gently into the bulkhead separating the hangar deck and the potato locker.

Mitchell, stunned but unhurt, stepped out of the burning cockpit. While he stood there, apparently just a bystander, one of the air group's chief petty officers asked him who had been in the airplane. The chief was astonished at Mitchell's answer, and rightly so. Mitchell began helping sailors caught in the fire from the crash, actually acting as a stretcher bearer for a man being taken to sick bay. In sick bay, after being treated for cuts from the Plexiglas in the canopy, Mitchell called the duty officer in his ready room. The officer almost fainted when Mitchell told him who he was. Soon, Commander Weisner and his operations officer were in sick bay, pleased to confirm that Mitchell was alive.[6]

Some inexperienced pilots had a delayed reaction to the stress of night flying in the Big Banshee. One superior pilot, who had earned terrific scores as a student pilot, went directly to VF-114 at NAS Miramar, California. When the squadron deployed with Big Banshees in 1956, he flew all the night practice intercepts and bombing missions for which he was scheduled, but after returning to the ready room after each landing, he went to the head and vomited. The pressure was too great for his system. This Naval Aviator wisely turned in his wings after being selected to be one of the first pilots in the next-generation night fighter.

■ ■ ■

The year 1955 saw the Navy complete development of other carrier improvements. The second most important new carrier system for improved night and all-weather operations was first tested when pilots used an optical landing aid—a mirror system, rather than the LSO's paddles—in the *Bennington* on 22 August 1955. Cdr. Bob Dose made the first mirror landing and Lt. Cdr. H. C. MacKnight followed two days later with the first carrier night landing performed using the mirror. The mirror and its successor, the Fresnel lens that followed the mirror into the fleet before the end of the decade, greatly decreased the number of carrier landing accidents, day and night, caused by the low-altitude paddles approach.

Optical landing aids had to be continually corrected for the carrier's motion, incorporating a stabilization system similar to those in fire-control systems. Improved stabilization systems also were incorporated into the ship-based, precision radar landing systems. As early as 1952, the Navy had begun development of a landing system that would ultimately be the Carrier-Controlled Approach (CCA) system. Test pilots from the Carrier Suitability Branch of the Navy's flight test center performed fully automatic landings on the *Antietam* using the Automatic Carrier Landing System (ACLS) with its SPN-10 radar, in the late 1950s.

The landing optical aid, or "mirror," on the *Bennington* as seen from an approaching airplane, 1955. *U.S. Navy*

The ACLS promised to be a great aid to night and all-weather aircraft at sea. For the next two decades, however, a human controller providing voice radio instructions to incoming pilots was the critical element of the precision CCA in carrier night and all-weather operations.

TACAN, a line-of-sight radio air navigation aid that indicated bearing and distance to a specific broadcasting unit, began development in 1948. ADF homers had given carrier pilots more precise bearing indication than the YE/YG. Now TACAN, installed in carriers by 1955, gave pilots distance measurements accurate to within one mile, as well as bearing information. In another breakthrough for tactical Naval Aviation, it was not long before an air-to-air TACAN system provided distance measurements between airplanes within line of sight.

In October 1955, the Navy commissioned the *Forrestal (CVA-59)*, the first carrier larger than the *Midway* class and the first built with an angled deck and steam catapults (tested by VC-3 and VC-4 pilots in 1954). As she began operating aircraft, the *Forrestal* also incorporated one of the first mirror landing aid systems.

The Navy had already begun turning the carriers' (now called CVAs, the new designation for the combat carriers) straight decks into angled decks. This modification incorporated not only the angled deck but also the steam catapult, the mirror landing system, and precision radars. After the modification, the *Essex*-class carriers were commonly called *Essex* (27C)-class carriers. Only five had been changed by the end of 1955, but twelve had an angled deck by the end of 1956 and all of the sixteen active CVAs (*Essex*- and *Midway*-class) had been modified by the end of 1957. During that period 85 percent of the CVAs made their next deployment with the angled deck incorporated. With the advent of the angled deck, mirror system, and steam catapults, carrier night operations became much easier and safer.

.   .   .

Because of the heavy Soviet submarine threat, several *Essex*-class carriers were converted to antisubmarine carriers (CVSs) to help locate and track the Soviet boats. Particularly in the Atlantic and Mediterranean theaters, Navy commanders assumed that the CVSs were high-value targets for Soviet bombers and needed twenty-four-hour fighter protection during their deployments at sea. By 1955, VC-4 began deploying VFN teams flying Corsairs on the CVSs in addition to the Banshees on the CVAs. In

May 1955, AD-5s equipped with the Korean War–vintage APS-19B air intercept radar replaced the F4U-5Ns. By November, VFN teams flying the ADs were deploying in the CVSs. Usually the VC-4 fighter pilots shared a ready room with the antisubmarine helicopter crews in the CVSs, putting a strange mixture of skills in the same space.

J. R. Flather, then a lieutenant, junior grade, recalls that the best part of flying in VC-4 at the time was the large amount of flight time an eager new pilot could get in the AD. Pilots averaged fifty hours per month. Quite a bit of their time was in actual instrument or night conditions over the ocean off NAS Atlantic City. Flather also recalls that although the flying with a CVS night fighter detachment was good, the old APS-19 radars hung on the AD-5s were a maintenance nightmare.

Because the Soviet air threat to the ASW carriers seemed less critical in the Pacific, VC-3 did not provide F4U-5N teams for CVSs. The squadron continued to deploy night Corsairs in Pacific Fleet CVAs until the F4U-5Ns made their last deployment in the *Princeton*, ending in April 1955. Later that year the Corsairs retired from both composite night fighter squadrons. By that time, almost half of the carrier fighter squadrons flew the Big Banshee, making further VC-3 and VC-4 teams unnecessary. VC-3's last VFN detachment returned to the United States in the *Shangri-La* (CV-38) on 2 July 1956.[7] By the end of that year, all the night fighters at sea in the Atlantic and Pacific Fleets belonged to the dual-mission Big Banshee squadrons, except for the AD-5N "fighters" in some CVSs.

In July 1956, VC-4 was redesignated VF(AW)-4, continuing to fly only the old AD-5s. Then that year, the squadron added another unusual item to its resume. A VF(AW)-4 team was assigned to the *Antietam* (now a CVS) to provide fighter cover during a planned round-the-world cruise. The commander-in-chief, Atlantic Fleet (CinCLantFlt), embarrassed to have propeller-driven fighters aboard the show carrier, decided that the team would fly the F9F-5, a day fighter. Some of the AD pilots finally received the jet experience that they wanted, but the team had no night flying during the cruise.

VF(AW)-4 continued to furnish Skyraider "fighter" detachments in the Atlantic Fleet CVSs through October 1957. Then suddenly the squadron began operating Banshees again after more than a year without jet experience. CinCLantFlt decided that the CVSs (the majority had a straight deck and no mirror) needed a high performance all-weather

94

fighter to meet the Soviet threat, so in November 1957, thirteen Big Banshees arrived on the squadron's parking ramp. FAWTULant trained the new jet night fighters for night intercepts at NAS Key West. That was the night training unit's last task, for on 30 June 1958 the Navy disestablished FAWTULant.

By September 1958, the first team of the new Banshee pilots was ready for carrier qualifications on the straight decks of the antisubmarine carriers. The attempt to conduct carrier qualifications for the F2H pilots in the *Tarawa* ended disastrously. A pilot approaching a bit too fast held his height above the deck after taking a cut signal, glided over the barriers, and crashed into two other Banshees on the catapults. There were no major injuries, but the incident ended the attempt to bring jet fighters back to straight decks.[8]

. . .

Before the Big Banshee had ever become operational as a night fighter, the Navy had already started new night fighter designs, leading to the Test Center's first flights of the new airplanes in 1951. There had been a design question of whether to put a radar operator in the new night fighters. The British, the Germans, and the Air Force had long ago decided that night fighters needed two crewmen. The Navy, however, had met its night fighter needs in World War II with single-piloted F6Fs and F4Us, and the F4U-5N still served the fleet as a front line night fighter after the Korean War. Moreover, the Navy had compared the performance of the two-seat F3D and the single-seat F2H, with the test results showing no significant difference in the ability of the two airplanes to conduct successful night intercepts of enemy airplanes. Therefore, the Navy had chosen to continue with single-seat night fighter designs, at least for the time being.

Jet engines were improving at such a rate that it was hard in those years to establish a new airplane design because a better engine was just around the corner. Still, a decision had to be made whether to install one or two engines in the new night fighters, taking advantage of the best engines available. The AJ and F3D night combat airplanes were the first operational multiengine carrier aircraft. They were followed by the Banshee and the A-3, which also had two engines, so there would have been no question of breaking a precedent if the next night fighter had two engines. However, that did not happen. A single, new-design engine could now provide a large increase in thrust compared to the two small engines in the F2H and F3D.

The first engine chosen by the airplane designers failed to meet its expectations after trials in a Douglas night fighter. Consequently, Douglas built the F4D Skyray around a different single, afterburning engine, giving the Skyray more than three times the thrust of the Banshee. The afterburner provided the extra thrust for takeoff, climb, and combat situations requiring maximum energy.

The Skyray set several speed and time-to-climb records early in the 1950s, but it did not get into operational squadrons for a few years. Because of its short endurance and sensitive carrier landing characteristics, it was not a popular carrier airplane. Furthermore, the engine burned so much fuel at high speed that the airplane had little endurance and a short range.

The F4D was an interesting airplane to fly because it was tailless and had no flaps. There is a story about a Test Pilot School student flying the Skyray for the first time. The flight manual and verbal instructions about the F4D had been skimpy. After a short flight exploring the flying qualities of the airplane, he was ready to practice slow-flight maneuvers in the landing configuration and make a normal landing at NAS Patuxent River, Maryland. Then the pilot discovered he could not find a flap handle in the cockpit. Believing that flaps were necessary to land safely, he searched and searched until he was in extremis because he was rapidly running out of fuel. He finally swallowed his embarrassment and called his duty officer, only to be told there were no flaps!

At first, the Skyray's armament was lacking in effectiveness. Four 20-millimeter cannon and a pod of 2.75-inch rockets that dispersed badly when fired were its primary air-to-air weapons. The rocket pod was soon replaced by the new infrared Sidewinder missile. The Sidewinder missile did not require the night interceptor pilot to get visual contact with its victim, but, unfortunately, the range of the missile was short and the search sector of the F4D's radar and missile's infrared sensor was small. Those factors meant that there was little difference between the tactics used by an air-intercept controller and F4D pilot team and those of the teams of FDOs and F6F pilots during World War II. In addition, because the four 20-millimeter cannons were the secondary weapons, any engagement finished with an attempt to get into a close position behind the enemy aircraft just as in earlier night fighters.

Following Captain Ramage's pilot transition plan, VC-3 trained pilots to fly the F4D as selected Banshee squadrons received the new night fighter. VF-213 was the first West Coast squadron using the Skyray aboard carriers. VF-101 performed the job of training new Skyray pilots at Key West for the Atlantic Fleet. In 1957, VF-74 became the first squadron flying the F4D on a deployment with the Sixth Fleet. The Navy ordered about four hundred Skyrays throughout the airplane's lifetime. It went to sea in eight Navy and several Marine Corps squadrons before phasing out of the carrier aircraft inventory in 1962.

. . .

The first model of the McDonnell F3H Demon, the F3H-1N, also failed its test program because its engine lacked sufficient thrust, but the designer soon installed the bigger J71 engine with an afterburner of about the same thrust as the Skyray's. The F3H continued to suffer growing pains along with the F4D. Some said the F3H was the best machine ever invented "for turning JP fuel into smoke and noise," but because the airplane had good carrier landing qualities, night fighter pilots preferred it to the F4D, its close cousin. However, the new engine had a unique problem: although the Navy classed the airplane as an all-weather fighter, the early J71s tended to quit in the rain. The engineers rather quickly found the cause, fixed the engine, and the Demon finished its career as a reasonable carrier night and all-weather fighter.

The F3H had the same practical speed limitation that the Skyray did—high subsonic. It had short endurance, about an hour and a half in the air before it landed. The F3H-2N carried four 20-millimeter cannon and four Sidewinders, just like the later F4D versions. The F3H-2M model, introduced in the late 1950s, was the first night fighter equipped with the radar-guided Sparrow I missile. The last model, the F3H-2, carried four Sparrow IIIs, two Sidewinders, and four cannon. In December 1958, Demon crews of VF-64 in the *Midway* and VF-193 in the *Bon Homme Richard* conducted the first firings of Sparrows outside the United States during WestPac deployments.

Automatic radar target acquisition (lock-on) and tracking was essential for the Sparrow system to operate effectively. The Demon's radar proved to be an excellent missile control system. In late 1960, during the *Coral Sea*'s deployment to the Western Pacific, VF-151 scored fourteen hits out of sixteen Sparrow firings at a towed Delmar radar-reflecting target. The record could have shown fifteen hits. One missile went ballistic when a pilot mistakenly thought he had locked on the Delmar's tow plane and shut down his radar, breaking the missile's control link and therefore causing an unnecessary miss.

Air-to-air tactics slowly changed to adapt to missiles that provided the potential to destroy enemy aircraft without seeing them visually, but the gun remained one of the night fighter's major weapons. With the Demon's reliable automatic tracking system, the F3H pilot received radar range and range rate information, providing good gunfire control solutions at night as well as missile-firing solutions. Fighter pilots still measured cannon-firing ranges in the hundreds of feet, requiring careful closure on the target from the rear quadrant. A good radar range rate indication made that maneuver easier and more effective.

Demon squadrons typically had poor flight time records, caused by Naval Aviation's maintenance and supply problems. VF-124, during the *Lexington*'s 1957 WestPac cruise, kept only two or three of their Demons in a "full mission capable" readiness state, that is, with all systems operable. With few flyable airplanes, flight time was minimal; at least one pilot reported only seventy-eight hours during the seven-month deployment and only seventeen night landings during his time in the *Lexington*, including four during the initial carrier qualification period. That equals about twelve flight hours and two night landings per month for the whole cruise, a bare minimum for proficiency in carrier flying.[9] Without a good, stable airplane and the *Lexington*'s angled deck and mirror for a landing aid, such little practice at night could have caused an extremely high accident rate.

Beginning in 1956, until VF-121 became the fighter replacement squadron on the West Coast, VC-3 also trained the first Pacific Fleet Demon pilots. VF-14 in the Atlantic Fleet received its first F3H in March 1956 and deployed to the Mediterranean in early 1958. The Navy operated about five hundred Demons throughout the airplane's lifetime. Fifteen squadrons flew the F3H aboard ship before the airplane's last deployment with VF-161 in 1964, two years after the Skyray left the fleet.

. . .

To improve the fuel—and therefore endurance—limitation of jet airplanes, in September 1955 the Navy set the policy that all future fighters would have aerial refueling equipment. The F3Hs and F4Ds had refueling probes compatible with the AJ tankers that had served the fleet since the middle 1950s. However, night refueling for Demon and Skyray pilots still presented the same problems of joining with the tanker and plugging into the tanker's refueling drogue that Banshee night fighter pilots experienced.

With the fleet introduction of the F4D and F3H almost simultaneously in 1956, the Navy deployed three types of night fighters through 1959: the Banshee, the Skyray, and the Demon. When the Banshees left the fleet in 1959, the Demons and Skyrays became the carrier's only night fighters. Squadrons were continually shifting from one airplane type to another. Because there were no two-seat F4Ds or F3Hs, the first checkout flights had to be successful. In those years the Navy's fighter community began to change from a group focused only on dogfighting to a group oriented toward precision instrument flying. The end result of the fighter pilot's maneuvering put his airplane in position to fire a radar-guided missile without ever seeing the enemy visually—day, night, or in clouds. Unfortunately, by 1957, although the Navy had modified all but two operational carriers with the angled deck and mirror landing aid system, accident rates among the night fighters remained high. Factors causing the high accident rate included the continual fluctuation of aircraft types and unreliable power plants and flight systems.

While the night fighters were moving into the jet era in the mid-1950s, VC-33 and VC-35 still flew ADs in VAN detachments on all CVA and CVB deployments, filling night attack, ASW, and ECM missions. Moreover, the VAN detachments deployed with two AD-4Bs for nuclear weapons missions in addition to their night attack airplanes, but they were beginning to lose that mission to the AD squadrons. By the middle of 1955, all the AD-4Bs belonged to light attack squadrons (after the advent of the heavy attack units, all the AD units were designated as "light attack" squadrons). However, the night attack pilots remained the only qualified AD nuclear strike pilots for about a year. Until the single-seat AD pilots qualified as nuclear strike pilots, the VC team pilots borrowed the AD-4Bs for the nuclear strike missions.

. . .

My personal experiences with carrier night aviation started during this period of the night composite squadrons' existence. I had met the night VC detachments' pilots while on my *Boxer* deployment during the Korean War. Night flying seemed to require the most skill of anything done in the air. As a student pilot in 1954, unaware that new aviators could go directly into night combat squadrons, I chose the antisubmarine syllabus flying TBMs because that course included more night flying than the fighter or attack aircraft courses. I planned to get into a night fighter or night attack squadron as soon as possible.

When I reported to ComNavAirPac (Commander, Naval Air, Pacific), I visited VC-35 hoping to see some of the old *Boxer* night flyers. VC-35, at that time the only night attack unit in the Pacific Fleet, had an impressive motto: We fly when sea gulls sit on the ramp! I convinced the operations officer, Lt. Cdr. Mark Hill, that I was highly motivated for night flying and would make a good night attack pilot. Hill and the ComNavAirPac training officer agreed to change my orders, and I reported to the Night Owls in June 1955.

My training in VC-35 was practically identical to the training given carrier night attack pilots since the beginning of the art. I was already familiar with the night flyers' night adaptation procedures, such as receiving preflight briefs in a ready room lit with red lights to preserve night vision. The offices and parking ramps were also lit with red lights or were not illuminated. Just walking to an airplane on the ramp at night was often confusing.

The AD-5N, which was replacing the AD-4N in the VAN teams at that time, was a better night airplane than the previous night AD versions because the pilot and one of the crewmen sat side-by-side, providing a better lookout and better communication between the two. That crewman operated a radar monitor and the passive electronic warfare equipment installed in the airplane. By stretching a little, the pilot could also see and interpret the radar picture on the scope. A second, usually senior, crewman, the radar operator controlling the APS-31, was more comfortable than in the AD-4N. He rode in the large rear section of the cockpit bubble, not in the belly of the airplane. Pilots relied on the radar operator for fine-tuning the radar and making the best interpretation of the radar picture. The AD-5N's instrument panel and lighting were the best of any 1950s night airplane. The panel had red edge lighting around the gauges and plenty of floodlights for backup illumination. However, the flashlight remained an essential piece of equipment for each crew-member. The AD-5N was slower by about twenty knots and could not fly quite as far as the single-seat ADs, but it carried an effective payload.

Before being assigned to my first detachment, I flew many practice instrument flights, two flights a day, daytime or night, in either the AD-5N or the new single-seat AD-6, reinforcing my confidence in the flight instruments' presentation. Trust in an airplane's instruments is absolutely essential for proficient night and all-weather flying. Some of the instrument flights were with a senior pilot observer in the second seat, but there were plenty of solo flights. On some of those, I was intro-duced to my future aircrewmen. The principal gyro instruments still pre-cessed and tumbled during maneuvering flight, limiting the degrees of pitch and roll that we did at night. Practice flying without the attitude indicator was a major part of the basic instrument training all pilots received in the night composite squadrons. Later in the syllabus, I began flying over water at minimum altitudes on nights with no moon and no horizon—just my instruments or my section leader acting as a horizon reference to keep me upright and out of the water.

An incident on my first cross-country flight in the AD-5N is typical of the confusion that could and did happen during flights in poor weather conditions. One weekend, I carried a fellow pilot to Denver for leave after a long deployment. Just before dark, we saw several thunderstorms over the New Mexican mountains, extending into Colorado. Night fell as we turned north toward Denver at Albuquerque. I was on an instrument clearance flying the old-fashioned, low-frequency radio airway, but was also using the automatic direction finding feature of the radio. However, because of storm induced static, I could not interpret the airway's "A" and "N" signals. Furthermore, the ADF needle was taking huge ninety-degree swings. I was unsure of our position. I was careful, however, to maintain a safe altitude over the mountains hidden by clouds and darkness.

Although my passenger did not have any control over the airplane, he could hear the radio and watch the instruments. In the worst of the storm, I looked over at him. He did not seem bothered. Because my friend had about twice my flying experience, I thought his calm demean-or indicated that we were on course. We broke out of the thunderstorms, ending the threatening segment of the flight, before we arrived at Denver. After my night landing at the unfamiliar airport, I told my friend that for a while I was totally lost and appreciated seeing that he was confident of our position. It turned out he could not follow the radio signals either and was just as lost as I! Better lucky than good.

. . .

In September 1955, I became a member of VC-35's Detachment H, scheduled to deploy in six months. The detachment had four airplanes and five aircrews—one pilot and two enlisted crewmen per crew—plus about thirty-five other maintenance men. We began the squadron's pre-deployment tactical training by practicing to join up in night rendez-vous, because we flew the majority of our missions as a section of two or a division of four airplanes. It was difficult to judge relative motion when the airplane on which one was joining showed only shielded exhaust stacks and dim lights. Joining on another airplane was much like the final stage of a fighter's night intercept on an enemy aircraft. I scared myself and my crew once by setting too fast a closure rate on the lead airplane. I saw the danger just in time to duck under and swish past the leader without a collision. I learned from the incident and never had a near-miss the rest of my career. However, my crewman that night did not accept another night flight with me until much later.

Our detachment made two deployments to El Centro for weapons training, one in November 1955 and the other not long before we deployed in 1956. We spent the weapons deployments flying one short flight each afternoon and two each night. During the first deployment we focused on dive bombing and close air support techniques. To ensure a hit while dive bombing, the pilot must place his airplane at the proper point in space with the gunsight "pipper" on the target. He must have the proper gunsight lead angle, dive angle, and airspeed. Finally, the pilot must make an estimated wind drift correction for airplane and bomb movement. Dive bombing, day or night, requires constant prac-tice to perfect.

Dive bombing at night, either with flares or by natural light, is more demanding because a pilot's eyes are not designed to find and track a tar-get in night lighting conditions. The art is doubly hard at night because there are no references for the pilot to check his airplane's attitude except the airplane instruments, the target, and sometimes a dim horizon. I learned why there was the difference of opinion reported during the

Korean War on the value of flares for night dive bombing. The flares created unusual shadows while lighting up the terrain and furnished a ghostly light, particularly on hazy nights. However, I decided early in my career that some light, even a diffused and shadowy light, was better than none if the target was not marked by its own light.

Laydown bomb delivery, which we also practiced, was a much simpler maneuver than dive bombing. Approaching the target while flying between 100 and 1,500 feet above the ground, the pilot released the bomb when the gunsight pipper or the nose of the airplane covered the target. Although flown at low altitude, the laydown was much easier than dive bombing. We planned to use that maneuver, day or night, to deliver the AD-5N's nuclear weapon or to attack ships or tunnels with conventional bombs.

We finished the week using live bombs practicing night CAS in a target impact area. Marine Corps forward air controllers (FACs), on the ground in positions where they could see the target, controlled our flights in daylight and at night. The team used flares to illuminate the targets during the night practice flights. I was expecting more from the experience of dropping live ordnance. I found that the smaller bombs did not affect the airplane at all. Dropping 2,000-pounders did result in a jolt as the bomb released and, if I flew too close to the explosion, the crew and I could feel the blast overpressure. I learned to compromise between getting close enough to the target to ensure a hit and staying far enough away to avoid the bomb's blast and fragment pattern.

The team lost a crewman and an airplane while over water off the California coast near San Clemente Island on a practice night bombing mission using flares to illuminate a target boat. The night was clear, but with no moon. We had completed about half the bombing runs when one airplane's engine failed. The pilot decided to ditch the airplane to keep his crew together in the water. Unfortunately, in the darkness he could not see the water and misjudged the airplane's impact point. The AD flipped in the choppy water, killing one of the crewmen. Both the pilot and the other crewman broke bones and could not get their raft out of the airplane. They would not have survived for long, unprotected at night in the cold December water, if not for the dramatic rescue effort that followed.

The Marine Corps's rescue unit at MCAS El Toro, about a hundred miles from the scene, heard the initial distress call from the flight leader. The unit immediately sent a helicopter out to sea to pick up our downed flyers. The skill of that Marine Corps crew was unbelievable because helicopters in those days had poor flight instruments and no stabilization system. Few people even flew helicopters overwater at night, much less executed rescue operations. Nevertheless, the helicopter crew arrived in about an hour, located the two men near a smoke light we had dropped, and picked both the pilot and the surviving crewman out of the water.

During the second deployment to El Centro, the team practiced the loft bombing maneuver used to deliver nuclear weapons. VC-35 pilots no longer routinely flew their host squadron's nuclear weapons aircraft on practice strike missions, but we did learn loft bombing, preparing for the unlikely event that we would be asked to fly one of the AD squadron's missions. Attack pilots developed loft bombing to throw a bomb some distance ahead of the delivery airplane while in a half Cuban-eight turn that allowed the airplane to reverse its course and get far enough from the bomb's burst point to escape the bomb's effects. We practiced loft bombing in AD-4Bs and AD-6s, because the AD-5N in which we flew night attack missions and ASW work was not fast enough to escape from a nuclear blast. Furthermore, because all of the ADs' attitude gyros were certain to tumble at the top of the half Cuban-eight—completely ruining the pilot's chance of maintaining a good attitude reference on a dark night—the weapon delivery segment of a nuclear strike would have to have been flown in daylight.

. . .

Before joining our air group, the team spent time on low-altitude navigation flights simulating nuclear weapons strike missions, about half of which were in the night hours. Other night radar navigation flights simulated the night interdiction missions that were common during the Korean War. Banning Pass, a gap in the southern end of the Sierras east of the Los Angeles Basin, was a favorite landmark for neophyte night crews. The pass has 8,000- to 12,000-foot mountains on either side, but is wide and level. It was narrow enough for the mountains to show on radar as we approached and flew through the pass. We stayed below the peaks, usually at about 5,000 feet, an altitude at which we needed the radar to ensure we did not run into the mountainside.

As we accumulated experience, we started exploring smaller, more exciting openings in the mountains to the east of the Pacific plain. The routes we followed included small country roads through the farm country that still existed in northern San Diego County and the Los Angeles area. We also stalked cars, trucks, and locomotives on the highways and railroads cutting across the desert by the Salton Sea and the Chocolate Mountains. However, even on the blackest nights, it was easy to find road traffic and trains because all the vehicles used bright lights. The training environment was not much like the dark landscape of an enemy country at war, as I learned later over Vietnam.

One VC-35 pilot who was between deployments started making a practice of carrying an ASW searchlight on the night road "recce" (reconnaissance) flights, checking out trucks on the desert highways. One night one of his target trucks, one that may have been illuminated on a previous flight, returned the brilliant light of his searchlight with a spotlight from the truck. The trucker's light hit the pilot right in the eyes as he began the pullout from his dive on the truck. Blinded by the light, the night pilot managed to pull out of the dive but returned his airplane to North Island with sagebrush caught on the tailhook.

The team members also practiced submarine localization techniques, including dropping sonobuoys at night. Placing a circular sonobuoy pattern in the water at night was extremely challenging. We flew the pattern as close to the water as possible, day or night, to reduce a buoy's drift as it dropped. After placing a center buoy, marked by a brightly burning smoke light, and while maintaining 50 feet or less above the water, the pilot began a thirty-degree banked turn to place four other buoys in a circle. In the bank, he had to judge the distance to the center buoy to set the proper circle radius. Invariably, while looking at the center light on a black night with no visible horizon references and his airplane tilted in the bank, the pilot seemed to be about 100 feet *below* that light—a strange and uncomfortable feeling.

Field carrier landing qualification in the AD, day and night, came next. Not all the CVAs had the mirror system, so the night pilots still learned the paddles pass. Because of the difficulty of the carrier night landing pass, we practiced hundreds of field landings at Brown Field and NAS Miramar before going aboard a carrier. The terrain around both fields was so hilly that in some parts of the carrier landing pattern, pilots flew below the tops of hills bordering the landing strips. At night that was disconcerting.

The night paddles pass had not changed since World War II. The pilot had to split his attention between his instruments in the cockpit and the world outside as he tried to see the ship's lights and the LSO in the darkness. The necessity to look outside the cockpit at night killed many carrier night aviators, not only attempting carrier landings but also in night bombing and ASW work.

We flew the downwind leg about 150 feet above the surface, then turned toward the groove when abeam the "carrier's" masthead light when it was visible or when the radio homer marked the abeam position. At night, I usually first saw the LSO's signals after completing about 135 degrees of turn toward the landing heading. Dressed in his suit with fluorescent stripes and using his illuminated paddles, the LSO indicated Fast, Slow, Low, High, High Dip (a bit too high), or Roger (speed and altitude OK)—then finally Cut or Wave Off. When picking up the LSO's signals so close to the ramp, I learned that I needed to be on speed, altitude, and attitude at that point because there was no time to respond to more than one signal. I usually saw a High Dip and a Cut. Because we flew the pass just above stall speed, when the engine power was cut the airplane fell to a three-point landing and, with luck, the tailhook caught an arresting wire.

Our LSOs sometimes used a different suit invented in early 1953 by a VC-4 LSO, Lt. K. C. Pailer. Rather than depend on black light to illuminate the LSO's stripes and paddles, this suit had strings of Christmas tree lights attached in vertical and horizontal lines on the suit. The lights continued around the paddles. The new suit allowed the LSO to move out of the ultraviolet light and relieved him of the necessity of wearing a shield for his eyes.[10] One pilot described the sight of the LSO at night:

> Among the problems was the proper illumination of the LSO. For reasons that are obvious, direct light would not work, so LSOs used black-light and fluorescent strips, causing no small amount of initial concern among the flight surgeons. Other attempts included a lighted suit, said to resemble a Christmas tree. The rig worked fine until the first time Paddles jumped into the net, wiping out the whole affair![11]

After perfecting my paddles pass, practicing to land on a straight deck, I had to learn different procedures for an angled deck and the mirror because our detachment received orders to deploy in the *Lexington,* a World War II carrier just modified with the 27C changes. The mirror approach started with the airplane at 300 feet, rather than 150 feet, above the water on a downwind heading. The pilot started the turn into the groove at that altitude behind the ship rather than abeam as in a paddles pass. The extra altitude added to the safety of the mirror landing compared to a paddles landing.

Landing on a straight deck, the pilot kept the throttles off after the cut and there was no stopping a crash into the barriers if the hook failed to catch a wire. When landing on an angled deck, however, there is a second chance, because there is no barrier blocking an immediate takeoff in case the hook skips over the arresting wires. For angled deck landings, the pilot had to learn to apply full power immediately after touchdown. That way, if the hook did not stop the plane, he simply took off again. The British, having invented the angled deck, also coined the term for a takeoff if the tailhook did not catch a wire, calling it a "bolter." I guess in the British imagination, the reaction of the airplane not catching a wire was like a man bolting through a door when caught in an undesirable situation.

After FCLPs, the detachment spent two weeks on the *Lexington* finishing our carrier qualifications. Neither the VC-35 LSOs nor the air group LSOs we joined in the *Lexington* yet trusted the mirror. The air group used paddles about half the time, day and night, during the following deployment, although the mirror and Fresnel lens landing aids eventually eliminated paddles passes. I had no trouble with my first night landings in the AD, but my anxiety was high. Lt. Max Puckett, one of our LSOs, explained to me (and probably to dozens of other wide-eyed nuggets) that every carrier night flight he flew had this pattern:

> Get up to the airplane in the pitch black. Finally settle in the cockpit, start the engine. Oh oh, that sounds bad, but engine checks OK. Guess I'll have to take that cat shot. Deck director answers my thumbs up and directs me to the cat. Last minute to change my mind. Full power, turn on the lights, the signal for the cat officer to fire the cat. Bang, rumble. I'm off the deck. What's that noise? Get on the gauges. Well, I'm OK—just fly the mission. About the middle of the flight, all's going well and the airplane is ticking like a clock. Back to the

carrier, into the landing pattern, downwind. Can't find the LSO—there he is. Work to get a "Roger" pass—the cut—bang on deck and the final jolt of the hook in the wire. Tell the maintenance chief waiting in the dark, Boy, that airplane's the best I ever flew!

. . .

Our detachment joined Air Task Group 1 for the deployment to the Western Pacific. Our detachment shared a ready room with the VC-11 AEW/ASW team and the AD squadron, VA-194. The detachments also received some administrative support from VA-194. The air group included two other night units: VF-52, flying F2H-3s, and an AJ team.

As well as having the angled deck and mirror, the *Lexington* had TACAN as her primary navigation aid. TACAN is much, much easier to use than earlier forms of radio navigation. Pilots follow a needle indicating direction to the ship's TACAN and watch the miles click off on the distance-to-station indicator. Wind drift compensation to keep on course is the only pilot skill required. At that time there were no special approaches used with the TACAN. Jet flights still returned overhead the ship, then executed a teardrop approach to the landing pattern. Our team's missions usually kept us under any overcast, so even at night or in low visibility we maintained formation until we made a downwind break into the final landing pattern.

As the cruise continued, the ship and the night flyers used the Carrier-Controlled Approach system more frequently. The *Lexington*'s CCA took control of an airplane flying at about 1,000 feet on the downwind leg of the landing pattern. The controller turned the airplane inbound when it was about six miles astern of the ship. Descending below the overcast, but no lower than 150 feet for paddles or 300 feet for a mirror approach, the CCA controller guided the airplane until within sight of the LSO or mirror. I have entries in my log book that show flights on which I logged several practice CCA approaches before landing.

The *Lexington* left for WestPac in May 1956. We stopped en route to conduct our Operational Readiness Inspection (ORI) in Hawaiian waters. The ORI's first exercise simulated nuclear weapons strikes on practice targets in Hawaii. Four of our five VAN crews, including mine, flew as pathfinders for single-seat AD pilots. As the ship approached the islands, we took off about 0200 one night about six hundred miles from a Hawaiian target that we planned to hit just after first light. The night was clear with some moonlight. Flying about 100 feet above the water with the AD-6 on my wing, the flight continued uneventfully for several hours. However, at the time my dead-reckoning navigation indicated that my radar operator should have detected at least one of the Hawaiian Islands, there was still nothing on the scope.

After about fifteen minutes, I decided to pop up to a higher altitude to get a better look with the radar, but still there was nothing. After another five minutes, I tried again. This time, the radar operator saw a return that indicated land near the starboard edge of his radar screen. I immediately turned toward it. The return kept getting stronger. It was Hawaii, the Big Island. We were about fifty miles off course. Without the radar, we would have missed the islands and conceivably could have run out of gas. I increased speed and corrected our heading for the target.

After the turn, I checked my gyrocompass against my magnetic compass, then compared compass headings with my wingman. My gyrocompass had precessed thirty degrees, putting us off the course to the islands. I should have been cross-checking the gyro with the magnetic compass, of course, because precession on long flights was common for gyrocompasses. As the dawn twilight increased, we passed Mauna Loa to port and arrived at the target on time for the AD pilot's scheduled weapon drop. When we landed and debriefed the flight to the ORI inspectors, I forgot to mention the minutes we were lost.

During the rest of the ORI, the team flew day and night bombing missions, doing excellent work hitting the Kahoolawe targets. An exception to that high-quality performance occurred on a flight I flew as wingman for a predawn simulated CAS mission. Each AD-5N carried its maximum bomb load of live bombs. There was no moon that night, with some low, scattered clouds and no horizon. Our instruments would be our only guide. The section leader took the first catapult shot and I was about thirty seconds behind him. I was safely airborne and raised my gear and flaps as soon as possible. My radarman could see my leader's blip on radar, but I could not see his dampened engine exhaust nor the faint fuselage formation lights that were the only lights we kept on over water away from other air traffic to give us a better appreciation of combat conditions.

I intended to stay in a trailing radar formation, but found myself closing rapidly on my leader and began to pull back on the throttle. With normal power settings on my engine, I still gained ground too fast. I slipped by the side of his airplane with my speed brake out, wondering what was wrong. I had little power on my airplane and my airspeed was down to near stall. When I called my leader, he said he was in trouble. His engine was at full power, but he could barely maintain airspeed and altitude. He jettisoned his bomb load on safe and told me his airplane was flying a bit better. I jettisoned my bombs also. The leader called the ship and told them of his emergency and that we were going to land at Barbers Point.

I closed into a wingman's parade position next to his airplane, but still could not see any details of his airplane. Then it started to get light. Imagine my leader's embarrassment when I told him his gear and flaps were still down, a condition that required the power he was carrying. When the gear and flaps came up, there was nothing wrong with his airplane! He had enough courage to admit his mistake when we returned to the ship at our scheduled time. I learned that even the best aviators have to be careful to keep up with basic procedures, particularly at night or in foul weather conditions.

The VC-35 and VC-11 detachments finished their ORI with an ASW exercise against a real submarine. Our crews had listened to tapes of submarine sounds on sonobuoys in training and understood the procedures for locating a submarine. However, although the exercise submarine stayed in a relatively small area for twenty-four hours, giving each of our crews a chance to locate and "destroy" it during day and night periods, we were not successful.

. . .

The *Lexington* steamed to Japan and spent a two-week period during June in Yokosuka Naval Base. We were at sea again, en route to Hong Kong, when the air group spent six days flying in the Formosa Strait between China and Taiwan. The air was still tense after recent incidents over the Quemoy and Matsu Islands near Communist China.

During July, our detachment heard that the squadron had been redesignated as VA(AW)-35. We had to change our report letterhead, but the name change did not affect our operations. We spent the rest of July in routine WestPac day and night flying. At night, the Banshees practiced night intercepts and did some night bombing. The AD squadron maintained their night landing qualifications and flew some overwater, night navigation flights, but the pilots did not practice night bombing. Our VC-35 team conducted night bombing practice using float lights as targets, flew night navigation flights, and practiced mock night attacks on the task group's ships. One night during the cruise we received permission from the ship's commanding officer to drop flares over the carrier and make dummy bomb runs. It added some realism to our training and gave the ship's gunnery department a different look at a possible air threat. We also practiced dropping night sonobuoy patterns, sometimes coordinating with the VC-11 antisubmarine hunters.

The LSOs closely monitored the day and night landing performance of all pilots in the air group. They kept a tally sheet with relative grades for every landing a pilot made. A five was an above average landing and a minus five was an FNKUA (F——— Near Killed Us All!) landing. I earned an FNKUA one night flying a paddles pass. I was OK, on speed and altitude coming up to the cut position, but because of moisture on the windshield, I could not see the LSO clearly. I unintentionally eased my airplane to the left while approaching the ramp, turning into the LSO and the port edge of the deck. I still could not see well in the dark and "forced a cut"—I was too close to the ship for the LSO to give me a wave-off, but in a hazardous position to land safely. The LSO gave a cut and I landed, catching the number two (of four) wire. The LSO said nothing on the radio.

I was the last airplane aboard that night, so my plane director on deck had me shut down the engine where I had stopped. When I stepped out of the airplane on the port side, my side, I almost fell into the water. It was so dark, I could not see that my airplane had stopped at the gunwales of the flight deck, just short of going overboard. The LSO was livid when he arrived in the ready room for the debrief. With that night landing I had nearly taken out the LSO platform as well as almost killing my crew and myself. I deserved the FNKUA score! However, learning from the experience, that was my only dangerous carrier landing in some 600 carrier landings, about 180 of them at night.

Months passed before a major night operation. At sea again in October, we were scheduled to participate in a big nuclear weapon strike exercise. Three carriers—the *Lexington, Bon Homme Richard,* and *Ticonderoga* (CVA-14)—participated in that operation. The carriers' launch positions for the exercise were in a triangle about seventy-five miles apart in the

A VC-35 Team Hotel AD-5N ready to catch a wire on the *Lexington,* 1956. *U.S. Navy*

Philippine Sea between the Philippines and Guam. Our practice targets on Okinawa were about seven hundred miles from the *Lexington* and there was a full-blown typhoon between the carriers and the targets! The admiral in charge decided to launch the strike airplanes because the weather was good at the carriers—just a heavy swell—and in the target area. The strike pilots had the hard job, navigating through the typhoon without getting lost or flying into the water. About two and a half hours of the eight-hour flights would be at night.

My regular radar operator had had his front teeth knocked out the night before we left San Diego. In his position at my back, the only way we could communicate was by intercom. Up to this time I had put up with the tongue-tied whistle caused by his lack of teeth, but the ship's dental lab had finally finished my crewman's bridge. I was glad that he could talk normally again, making our conversation easier on this night flight in a typhoon.

Again I was a pathfinder for an AD-6 delivery pilot, who decided to lead our section with me assisting with the APS-31 when we approached landfall. About 0300, he was in the first airplane launched by the catapult crews, with me right behind. I joined in a trailing radar formation and we were off into the dark. We stayed at about 100 feet, measured by the radar altimeter, under low clouds no more than a few hundred feet above us. There was enough visibility to see my leader's bright lights, but it was *black*. After about an hour, we broke into the eye of the typhoon. Inside the eye, the air was perfectly calm and starlit, but then we entered the storm again. Daylight started to break and I tensed as I became able to see below me. The waves seemed to be lapping at our airplanes even though we were 100 feet above the water. The sea was *rough!* We broke out of the typhoon about an hour and a half away from Okinawa.

My leader and I checked our dead-reckoning estimates and compass headings and agreed on our estimated position. This time my radar operator picked up landfall at Okinawa on time and at the proper position. The AD-6 pilot made a beautiful weapon drop and we went home, this time flying a little higher through the storm, still in radar formation with my leader flying on instruments through the rain.

The *Lexington* was ready to recover us on a "ready deck" when we reported that we were fifty miles away. The mirror approach was tough because of the pitching deck in the swell. The mirror's "ball" jumped almost instantaneously from indicating a high to a low because the stabilization system could not keep up with the ship's movement that day. Aided by the LSO, I had to estimate where the proper glide slope was located. As always, the LSO had the responsibility of giving a cut or wave-off signal. His judgment was most important with a pitching deck to keep an airplane from hitting the deck just as it was on an upswing. The conditions were the worst in which I have tried to land on a carrier's deck.

We left for the United States in the first week of December 1956. The air group's night operations record for the cruise was typical of Pacific Fleet air groups during the middle 1950s. Deployed carriers had increased their night operations to maintain the proficiency of more than half of the squadrons: the F2Hs, ADs, and AJs, as well as the VC units. Carrier flag officers and captains began changing the ship's operating hours: sometimes noon to midnight or midnight to noon, sometimes all night, sometimes all day, to prevent the flight deck crews from working twenty- to twenty-four-hour days but still keeping the night flyers qualified. During the late 1950s, a Banshee or light attack pilot could expect about fifty night flight hours and twelve carrier night landings during a carrier cruise in either the Atlantic or the Pacific Fleet. My record on the cruise was typical of VC-35 pilots during that period. I had over one hundred carrier landings in the *Lexington,* about twenty-five at night—quite an accomplishment in those days.

. . .

It was customary for VC-35 pilots between deployments to act as instructors for aircrews getting ready to go to sea. I taught new squadron pilots about the AD-5N's weapon system and nuclear weapons. I worked with, among others, Lt. Jack Shaw, who was one of the growing group of Naval Aviators who also were electronic engineers. These men advanced and supported the fleet's technical knowledge about radar and the electronically controlled weapon's systems that were in a large part responsible for the growing effectiveness of night and all-weather squadrons. Shaw's knowledge of electronics and creative drive allowed him not only to train other pilots and aircrewmen but also to improve circuitry in the radars and other systems in the night ADs. As an example, he designed a prototype radar circuit that displayed an adjustable range line on the radar scope. That enabled the radar operator and pilot to see with precision what their range was from a target or other radar landmark. All the VC-35 AD-5Ns received this modification. Despite his engineering skills, Shaw was first and foremost a Naval Aviator. He was killed about ten years later when an arresting wire broke during his last carrier landing as commanding officer of VA-125.

Because of a series of fatal accidents in the squadron, my home squadron tour was cut short as deploying teams were reorganized. I became the operations officer of Detachment J. In March 1957, the team went to El Centro for new and different nuclear weapons work. A group at NAS China Lake, California, had developed a new nuclear weapon especially for night and all-weather delivery by the AD-5N. They had taken the Tiny Tim aircraft rocket motor and put the warhead of a nuclear weapon on its front end. China Lake had named the weapon BOAR—"Big Old Aircraft Rocket." Because of the added distance separating the delivery airplane and the bomb burst provided by the rocket assist, AD-5Ns could escape the blast of the nuclear weapon. There was a slightly different delivery maneuver also, a ninety-degree wingover turn after weapon release rather than a half Cuban-eight. The wingover was much more comfortable at night or on instruments, particularly because the attitude gyro did not tumble in the maneuver. We now had a second, more versatile nuclear weapon that the AD-5N crews could deliver at night and in any weather condition.

Pilots and radar operators in VC-35, VC-33, and the heavy attack squadrons were the only people in the Navy using the output of the intelligence groups that were collecting radar scope photographs. The heavy attack and night attack crewmen made a strong contribution to the art of radar scope photography. With little radar schooling, radar interpretation was for them almost a self-taught art. While working in the home squadron, one of the VC-35 aircrewmen's tasks was to assist those in the Navy's air intelligence branch who built libraries of radar scope photography corresponding to the files of optical target photography. Besides actually flying over likely targets and taking photographs of the radar scope, duplicating the probable scene on a radar could be accomplished in detail by creating a scale model of the target terrain and the facilities at a target. The first terrain models simulating a radar bombing target that I saw at North Island were primitive, but when taken from the proper angle, pictures of even those early models gave a reasonable approximation of a target's actual radar image.

While training for deployment, night attack crews tried to find terrain and facilities that were similar to possible bombing targets, so that the radar operators could see a scene resembling an actual target. During the training cycle before my 1957 deployment, I searched the charts of southern California looking for locations that matched the landfall point and aimpoint of the principal target I was to attack if we went to war with the Soviets, which was an airfield. There was a small airport at what is now John Wayne Airport in Orange County. My crew and I spent several hours, during the day and then at night, flying approaches from the ocean and up a gorge to our simulated target airport. We began to understand what we might see on the radar scope and outside our airplane if ordered to strike. One environmental feature existing in Russia during part of the year that we could not duplicate created a problem for us: We could not get radar scope pictures with snow covering the land. There was not much snow in coastal Southern California.

. . .

Detachment J joined its air group and spent June at sea off California working with the *Kearsarge* (CVA-33), our new home, before the WestPac cruise. Unfortunately, one of our pilots and his crew were killed one night attempting a rolling takeoff in their AD-5N while carrying a maximum fuel and bomb load. When not flying, it was my responsibility to check the length of deck run required for a launching aircraft's weight and the existing wind-over-the-deck. In our ready room before that fatal flight, I had checked the calculations of the pilots and determined the deck-run distance that the AD-5Ns needed for the wind-over-the-deck that the *Kearsarge*'s skipper promised. Everything was in order.

When I arrived in Pri-fly to check the launch, the flight deck officer had already put our first airplane in position and the pilot was checking his engine for takeoff. I immediately saw that the airplane needed fifty feet more deck run for a safe takeoff. Somehow in haste or through inexperience, the pilot had not noticed the fifty-foot difference between his takeoff spot and the spot he had calculated. I told the air officer, a commander who was in charge of the launch, that the distance for takeoff was too short. He did not pay attention to my loud requests to hold the launch and correct the AD-5N's spot, saying that it did not make any difference. Besides, to do that would have meant a lot of work respotting airplanes to get the added room. The other alternative was a catapult shot, but the air boss was not ready for that either. When the flight deck officer released the airplane for its night takeoff, the pilot managed to get off the deck, but immediately settled into the water in front of the ship. I believe the *Kearsarge* ran over the airplane and crew. Search crews found no wreckage or survivors that night or the next day.

Later in the cruise, while steaming southeast of Okinawa on a stormy day with no flight operations, a destroyer transferred a man by highline to the *Kearsarge*. The man had a perforated ulcer and was bleeding to death. The destroyer's chief corpsman could do nothing further, so his ship transferred the sick man to the *Kearsarge* where it was thought the patient could get more help. It turned out that the *Kearsarge* doctors could not help either. They suggested that the patient be flown ashore as soon as possible. By then, it was about midnight. The weather was still terrible, with high winds and heavy rain as well as strong swells. At times green water was breaking over the bow.

The ship's skipper and the commander of the air group (CAG) agreed that VC-35's night flyers should perform the medical evacuation flight that the doctors wanted. One reason that the CAG and the captain chose our team was that we had an airplane with space to carry a stretcher and a doctor. Second, we were the best-trained and most experienced night and all-weather flyers in the air group. The air group commander called our officer-in-charge, who told me to get a crew ready to go. I picked my radar operator, who would be my only crewman because the senior flight surgeon aboard, a lieutenant commander with some flight experience, chose to accompany the patient.

The aerologist forecast bad weather all along the five-hundred-mile route to Atsugi. The air operations officer told me that he had filed my flight plan with the Japanese and that NAS Atsugi GCA would be ready to pick me up in about three hours. I had my radar operator set up the radar in the rear cockpit while we were still on deck, before he climbed into the right seat of the AD-5N. The doctor and his patient strapped into the rear section of the cockpit bubble.

The ship launched us at about three in the morning. Having decided that I could navigate better by checking the wind signs on the water after dawn broke, I stayed below the clouds, but we were still in rain. It was a quiet flight as I concentrated on flying on instruments through the darkness and rain squalls. When dawn came, our radar showed the coast of Sugami Wan at a point on the bay's shore south of NAS Atsugi, just where I had planned. As I climbed into the soup, there was some St. Elmo's fire—the first I had ever seen—flowing from the wings.

I called Atsugi's GCA, and I was relieved when they answered. The GCA crew did an excellent job guiding me on a smooth approach in the pouring rain. I had to go down to the instrument approach minimum altitude—probably below minimum—before I saw the runway and landed. The runway and taxiway were rivers, but I managed to steer the AD up to the director on the ramp in front of a hangar. Then an unusual thing happened as the lineman directed me to taxi right into the hangar. That was usually not done, but it was raining so hard, the Atsugi doctors had asked the ground crew to bring me inside before I shut down the engine. The patient stayed dry as the corpsmen lifted him out of my airplane.

I still am proud of my only chance to save someone's life. Without the special night and all-weather training our team had, the patient would not have survived. The flight surgeon told me after the flight that his patient could not have lived five hours more with only the equipment on the *Kearsarge*. We returned to our carrier the next day in much better weather. The front had gone by and the ship was nearer Japan, so our trip was a pleasant one hour's flight.

. . .

In the early part of the winter that year, the ship spent two weeks in Subic Bay in the Philippines. The air group put about half the airplanes ashore at NAS Cubi Point to maintain the pilot's flight proficiency. While at Cubi, each of our team flew a night training mission on the profile of our planned nuclear weapon strike missions, all of which were maximum-range flights, to check the AD-5N's fuel flow figures. Completing the flights also checked our ability to fly a realistic night mission over unknown territory.

I selected Davao on Mindanao as my simulated target because it would present a radar picture similar to my Russian target. My crew and I left Cubi at about 1700, planning to be at the target at 2100. Sunset was about 1830. I estimated our landing time to be 0100 the next morning. The route took us over and around the Philippine Islands between Luzon and Mindanao. The final leg of the flight to Davao was down in a straight river valley between tall mountains. As I skimmed over the two-hundred-foot jungle treetops in the moonlight, I felt a sensation that I knew later in combat. It was exhilarating, flying in conditions in which few people had done or will do.

The final run toward Davao airfield was at a measured 100 feet over a plain of rice fields. My radar operator found the blip of the airfield's hangars and measured the distance from the field to our initial point. By that time, the throttle was fully forward as the fat AD slowly reached its maximum speed. When the radar operator called "initial point," I punched a timer and waited a few seconds to put on four g's. When the airplane pitched up about thirty degrees, the weapon system indicated that it had released the simulated weapon. Although there was moonlight, I was careful to watch the attitude indicator and altimeter as I rolled into a 100-degree bank to reverse course. Descending to 100 feet again after finishing the turn, the escape maneuver had been completed. The flight back to Cubi was a sightseeing trip in the moonlight shining off the tropical beaches. We landed, tired but pleased with the flight and the accuracy of our fuel planning. All the team's crews reported successful flights.

During the at-sea period following the Cubi visit, the Banshee squadron skipper and one of his junior pilots flew to Japan to pick up a Banshee radar part. On the return flight, they misestimated the wind speed aloft and ran out of fuel about one hundred miles from the ship. They had been talking to the *Kearsarge*'s air controllers, but because all the AJ tankers were out of commission, there was nothing that could be done. At that time there was still an argument about whether it was better to ditch or eject from a jet, but the final decision in an emergency was up to the pilot. In this case, the junior pilot ejected and climbed into his raft safely. The skipper decided to ditch and was not recovered.

The destroyers escorting the *Kearsarge* raced to the estimated position of the downed pilots, but unfortunately the destroyers did not find either pilot. We had immediately started to plan an air search to begin the next day. Our detachment and the ADs flew four-hour sector-search missions the next morning and afternoon. Our AD-5Ns continued the search at night with one four-hour flight. The search continued in the same cycle the next day, but still the searches found no sign of the pilots. But on the second night, using the ASW searchlight, our officer-in-charge's flight found one pilot! Seeing anything as small as a man in a raft at night is a miracle, even with a searchlight. The wind and current had blown the downed pilot far from the first estimated position. The flight dropped smoke lights in the water to keep track of the pilot as a destroyer closed on the spot at the edge of our square search pattern. At first light, the destroyer recovered the junior pilot. He was returned to the *Kearsarge* later in the day to tell his tale.

We next participated in an unusual, cold weather strike operation named "Castle Rock." During the exercise, the strike pilots flew their planned nuclear strike flight routes up to landfall at the Russian coast before returning home. Air Force fighter-bombers based in Japan acted as an opposing force attempting to find and attack the carrier. Commander, Task Force 77 Rear Adm. Fitzhugh Lee chose to keep the carrier close to the eastern shoreline of Japan, relying on the radar shadow of the land to prevent detection by "enemy" aircraft. The carrier planned to stay in full radio and radar silence during the two-day exercise, forcing returning aircrews to use their best dead-reckoning navigation skills to get back to the ship. However, the weather turned sour, with snow, ice, and low visibility, and the admiral shortened the exercise to one day and launched only a few of the AD-6s and the VC-35 team.

Detachment J launched two sections at about 0300 into the snow-storm. Our officer-in-charge led the first section off the deck. After my catapult launch, my wingman shot off the deck and joined me in radar formation. Our route passed through the Tsugaru Strait between Hon-shu and Hokkaido, a part of Japanese waters over which I had never flown. I went through the strait that night navigating through the snow with the aid of our radar. The images of the mountains on each coast showed clearly on the radar scope. I flew at 200 feet above the water, measured by the radar altimeter, but was not able to see the water's sur-face because of the snow. Because of our training, we were comfortable and in no particular danger—unless, of course, the engine quit.

I was on schedule when we started across the Sea of Japan toward my planned target, an airfield near Vladivostok. The snow was still falling, but luckily there was no ice forming on our aircraft. The engine ran noi-sily, but OK. Because of the horrible weather, if my wingman, his crew, and my aircrewmen had not been watching me, I would have been tempted to turn around for the kinder waters of coastal Japan. While still flying in those conditions, our planned landfall on the Russian coast began to show as green blips on the radar screen. I continued on, with some intention of flying over Russia, but then, about five miles offshore, I had second thoughts. The mountains ahead were about 5,000 feet high. We would almost certainly have encountered icing conditions in the snow at a higher altitude. I decided that going this far toward Russia without being intercepted was exciting enough and turned around.

After returning across half the Sea of Japan, my crew and I started to relax. I stayed at sea level on the return trip to avoid ice on our ADs. The snowstorm ended when we passed through the Tsugaru Strait and entered the Pacific Ocean. With the *Kearsarge* in radio and radar silence, we found her using our radar and landed without incident.

The ship headed for the next port call, Hong Kong, down the Formosa Strait, where the admiral's staff had scheduled another strike exercise against the Chinese Nationalists on Taiwan. On the first night of the exercise, I led a four-plane simulated bombing raid into Keelung, Tai-pei's harbor. Rather than risk detection by the "enemy" over land, I led the group in a trailing radar formation through the harbor entrance, which was about as wide as San Francisco's Golden Gate. After entering the harbor, the formation spread on different headings toward different sectors of the port. The excellent radar work by our crewmen ensured that if the mission had been an actual attack, our hits on the stationary ships would have been devastating.

When we returned to the ready room, Capt. Bill Martin, at that time Rear Admiral Lee's chief of staff, sat in the front row. He said he wanted to listen to our flight's debrief. Knowing that he had been a night bombing pioneer, I thought it was an honor to have such an experienced Naval Aviator and night flyer review our work. I described our attack plan and the flight, before each pilot gave his comments concerning what had happened that night. Then Martin told us about his night raid into Taipei's harbor during World War II. Although his plan and mine were almost identical, the results of the two flights were different. Martin's strike sank four Japanese ships but lost three airplanes to antiaircraft gunfire. In contrast, our team had flown only a realistic and exciting, but peaceful, training mission.

Our last WestPac operation, in late February 1958, was Exercise Strong Back, a Seventh Fleet amphibious, conventional-weapons strike and air defense exercise. It was set up to test the Seventh Fleet's amphibious force's ability to move several hundred miles and conduct an assault landing while threatened by enemy air. Our air group's mission was to locate and "destroy" the amphibious force.

Our officer-in-charge had gone on emergency leave, leaving the team with four airplanes, four pilots, and five sets of aircrewmen who were scheduled every day of the exercise for four four-hour sorties in the morning and four four-hour sorties at night. Our maintenance crews kept the airplanes flying, and we did not miss a sortie during the exercise. Each VC-35 pilot flew over thirty-two hours in four days, half the time at night. Today's jet pilot has worked hard if he flies thirty hours in thirty days.

I believe that the group's Banshee squadron located the amphibious ships first on the initial search mission, but only because they were faster and covered the sea quicker than our ADs did. Our flight also sighted the amphibious force on the first morning and conducted mock attacks. Each strike flight was a little different because the amphibious ships kept moving until they ended up in the Amphibious Operating Area (AOA) off Linguyan Bay. We kept changing our attack tactics, approaching sometimes at night from the land, at other times from the sea. The amphibious force had no air support, so it was hard to tell how we would have succeeded during a war.

The last day and a half required long flights to get from the *Kearsarge* to the AOA for our mock attacks. I was extremely tired when returning on my final night mission during that exercise. I had led the team's four aircraft to the AOA, but we had split up for our attacks and returned to the ship singly. The *Kearsarge* was still in radar and radio silence, so we and the night Banshees relied on aircraft radar to find the ship. It was about midnight when my radar failed. I estimated the *Kearsarge*'s position and started a square search to find her. After some time, the ship called me. I was the last plane out and they had been watching me on their scopes, but had continued the exercise to the last minute. I was just about ready to break radio silence when the ship called. After the ship told us in which direction to look, we saw the dim lights of the *Kearsarge*. All of my crew were so tired, we just did not see the lights without a cue as to where to look. I made a good landing to end the exercise. The incidents I have recounted about my experiences in VC-35 were typical of the flying done by the night attack detachments on peacetime deployments in the late 1950s.

. . .

During the final years in which the night attack squadrons operated, electronic countermeasures tasks became more important to the carrier task forces. The Douglas Aircraft Company responded by modifying the AD yet again to produce the AD-5Q. That version was configured not as an attack airplane but to carry the latest passive and active electronic warfare systems. Although the "Queen's" active systems did not have much power, the airplane was a formidable carrier asset. Teams from VC-33 and VC-35 deployed in 1958 and 1959 with at least one AD-5Q in the detachments.

The last moment in the death of the night air combat composite squadrons finally arrived. VC-3 had stopped sending night fighters to sea in 1956. Three years later, the last VA(AW)-33 and VA(AW)-35 night attack detachments deployed. That year, in July 1959, the Navy disestablished both squadrons. Hanging on for almost another year after the flight deck catastrophe in the *Tarawa*, but with little money to sustain operations, VF(AW)-4 was also disestablished during the summer. That act closed down the night air combat composite squadrons. Over the ten years of those four squadrons' existence, carrier aviation saw tremendous advances in systems that eventually allowed all carrier aviators to fly at night and in any weather condition.

As the Big Banshee was losing its place as a night fighter to the Skyrays and Demons, the F9F-8B Cougar and F7U-3 Cutlass began to displace the Banshee as a jet nuclear and conventional weapons bomber. Because it caused so many fatalities, the Cutlass made only four or five deployments as a carrier airplane, lasting only from 1956 to its last deployment with VA-66 in 1957. The Cougar bomber version, however, was basically a good airplane. Its predecessor, the straight wing Panther, had been a workhorse during the Korean War as a day fighter-bomber. After being wired for nuclear weapons, the F9F-8B served for a time as the first successful jet light attack airplane and necessarily began night operations. Besides having nuclear weapons systems, it was also a good dive bomber. Pilots of VA-26 flew the Cougar bomber to win the air-to-surface events during the Navy's 1956 weapons meet. It was not capable of long range flights, however, so the F9F-8B was only an interim solution to the need for a good jet light attack airplane. The F9F-8B squadrons flew the minimum number of night hours and made the minimum number of carrier night landings to maintain the carrier night qualification necessary for nuclear weapons pilots.

The A-4A Skyhawk joined the fleet as the next light attack airplane in September 1956 (although a bomber version of the FJ entered the fleet at about the same time as a light attack airplane, it did not last long in the fleet). The A-4 was an austere airplane designed only for carrying one nuclear bomb a long distance. VA-72 in the Atlantic Fleet was the first squadron to receive the A-4A. Lt. (jg) Jerry Zacharias was in the squadron and remembers it as a good, stable airplane. About the time Zacharias moved to another duty station, the squadron was beginning night landing practice.[12]

The A-4A and A-4B had little inherent capability for the night operations in which its pilots had to participate. The instrument panel was as bare as a day F6F's panel in 1944. The airplane had only the basic instruments for flight in foul weather and for carrier night flight operations. The other night airplanes—ADs, F2Hs, F3Hs, F4Ds, and AJs—all had radar altimeters; the early A-4s had none. Ready for launch on a night mission, the A-4 pilots often heard a Demon pilot "down" his airplane—declare it unflyable—because the radar altimeter was bad. The "Scooter" pilots just laughed and went into the blackness of night with only their confidence in the old barometric altimeter to help them stay out of the water or off the mountainside. Flying only nuclear weapon strike missions, it was not until the early 1960s that the A-4 squadrons began practicing night dive bombing and night interdiction missions.

A new and better heavy attack bomber arrived in the fleet when HATWING-1 received the twin-jet A-3 Skywarrior in March 1956 as a replacement for the AJs. The first A-3s went to newly established VAH-1 at NAS Sanford. It was a large airplane that could carry conventional bombs, as well as several versions of nuclear weapons. The all-weather A-3 carried a pilot, a bombardier/navigator (B/N), and a flight engineer. Although large and heavy for carrier operations, it was a stable airplane with good flying characteristics.

In contrast to the AJ program's pilot selection process, most of the A-3 pilots had previous carrier experience. Consequently, most of these men had plenty of carrier landings, but most had little night flight time. Besides becoming familiar with night work, flying with a crew was also something new for the experienced fighter pilots transferring to heavy attack. Aircrew manning for the A-3 presented another problem, as selection of radar operators for the heavy attack units became more difficult. The supply of World War II and Korean War radiomen and avionics technicians who had flown in the bellies of TBMs and ADs was leaving the Navy. In the late 1950s, the Navy began commissioning the bombardier/navigators, encouraging more men to volunteer as B/Ns. This program was successful, bringing in more educated men for not only the sophisticated systems in the modern heavy attack airplanes but also the next-generation night fighter.

When the A-3 first deployed, there were five large *Midway*- or *Forrestal*-class attack carriers, of which three had angled decks. The advent of the A-3 made the largest *Forrestal*-class carriers even more advantageous because of their size. By the time the A-3s deployed, all carriers had a mirror landing aid system, so the A-3 pilots did not have to learn the paddles pass. The first A-3 squadrons deployed in the *Forrestal* and other Atlantic Fleet carriers in 1957, displacing the AJs.

For the first Pacific Fleet deployment of the A-3, the *Midway* and VAH-8 went to the Western Pacific in early 1958. The A-3 squadrons joined the ADs, Banshees, and A-4s as the sea-based nuclear weapons airplanes. The aircraft's long range and all-weather systems made it valuable to the fleet commanders charged with the nuclear strike mission. By 1959, A-3s deployed in the *Ranger* (CVA-61) and *Midway,* adding to the Pacific Fleet's nuclear weapons strike posture. The number of A-3 squadrons in the Atlantic Fleet had increased to six by the end of the decade, and there were five heavy attack squadrons flying A-3s in the Pacific Fleet.

With the deployment of the A-3s, all of the improved AJ-2s became tankers, supplying fuel to airborne jet fighter and attack airplanes. Full AJ squadrons continued to deploy in the *Midway*-class carriers and three- to four-airplane detachments deployed on the *Essex* (27C)-class carriers through 1958. The Savage aircrews of a VC-7 detachment in the *Essex* helped end a Formosa Strait crisis that year, but by 1959 no AJs deployed. The heavy attack, all-weather, nuclear bombing mission rested with the A-3 squadrons. Furthermore, the new A-4 "buddy" tankers became the only aerial refuelers in the carriers.

Chapter 6 **A NEW ERA FOR CARRIER NIGHT AND
ALL-WEATHER OPERATIONS**

Setting an exact time for the beginning of a new era for carrier night and all-weather operations is sure to be imprecise, but certainly by 1960, carrier operations were different than in prior years. Between 1954 and 1960, shipyards finished the carrier modification incorporating the angled deck, steam catapults, visual landing aids, and improved navigation and approach systems in all CVAs. Furthermore, the new large *Forrestal*-class carriers gave the night and all-weather aviators a more stable flight deck in rough seas and had wider landing areas than the older carriers.

By 1960, all squadrons except one day fighter squadron in each carrier air group had a night and all-weather mission. Before 1960, a carrier launched at most four night fighter and attack sorties per night cycle; in the new era, the three to four squadrons with a night mission could fly between eighteen and twenty-four sorties per night cycle. The night air combat potential of the carriers increased four to six times. Night flying finally had a higher priority in the carrier's daily activities than did the all-hands hangar deck movie.

In the early 1960s, the Navy took another leap forward affecting carrier night operations. The commanders decided that the carriers could light up their flight decks with bright overhead floodlights. The floodlights lit up the flight deck as if it were daylight. The air bosses seldom used white floodlights because red or yellow "moon" lights gave almost as much illumination for night deck operations and seemed to give fewer shadows than white lights. No longer would flight deck crews operate in complete darkness. Along with the floodlights, the carriers began to get brighter yellow or white angled-deck centerline and deck-edge lights. The ships added "drop" lights—a lighted, vertical line from the flight deck to the waterline that marked the center of the ramp on the carrier's fantail. No longer would pilots have only dim centerline and deck-edge lights to mark the landing area.

Improvements to the SPN-10 radar for precision instrument approaches continued into the 1960s. Tested on the *Midway* by Lt. Cdrs. R. K. Billings and R. S. Chew, Jr., flying the new F-4, by June 1963 they had ended six months of work on an improved ACLS, a system that was still primitive compared to a modern ACLS. At that time, few airplanes had the electronics to use the ACLS for landings. Ships that had the SPN-10 installed used it only as the final approach radar for CCA controllers' talk-down approaches.

Taking advantage of the improvements in navigation aids and approach radars, in the early 1960s the carriers started using a straight-in TACAN approach (see figure 4) similar to the high-altitude, standard instrument approach to an airfield. The overhead teardrop pattern for the jets' night and all-weather approaches remained as an alternative approach if wind conditions or limitations to the carrier's maneuvering room made a straight-in approach unfeasible.

Before each launch during inclement weather and at night, the carrier designated a "marshal" position described by a TACAN range and bearing from the ship. The marshal was on a bearing downwind from the carrier's predicted course for flight operations. If marshal changed while a flight was airborne, a controller gave the flight leader the new position when he returned. When separate flights returned to marshal, each received instructions to hold and maintain different altitudes at marshal until ordered to begin the final landing approach. The fighters usually were first to start the approach, followed in order by the light attack, heavy attack, and support airplanes.

When their "marshal time" occurred, the flights broke up at two-minute intervals, letting down individually to the carrier-controlled approach that had become standard for operations at night or in instrument conditions. Leveling off at 1,200 feet and ten miles from the ramp, each pilot slowed to final approach speed and lowered his landing gear, flaps, and tailhook. When six miles from the ship, the pilots descended to 600 feet. At one-and-a-quarter miles from the ship, guided by the con-

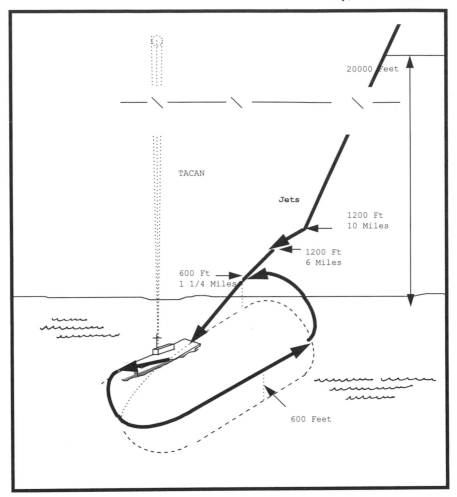

Figure 4. TACAN approach to mirror landing pattern.

troller's instructions, each pilot descended until he saw the "ball" in the lens or reached ceiling or visibility minimums. The pilot then landed using the visual landing aid indications or waved off for another approach.

. . .

In 1957, day fighter squadrons had begun receiving the supersonic F-8 Crusader. The Crusader's superior performance made the night fighter pilots—who were still flying Banshees, F3Hs, and F4Ds—drool. The

Navy saw the promise of a Crusader night fighter, and soon design modifications appeared to fulfill that promise.

Paralleling the night F-8's development, the F-4 Phantom II, a true night fighter, flew first in 1958. The first plans for the F-4 showed it as an attack airplane, but quite soon it became a twin-engine night fighter, beating another version of the Crusader, the F8U-3N, in a head-to-head competition for the role as the Navy's next all-weather interceptor. The Phantom flew controlled by a pilot and a radar intercept officer (RIO) who operated the AI radar. Cdr. Jerry O'Rourke, among others, had been advocating twin engines and a two-man crew for night fighters ever since his combat tour with the F3D in Korea. There were still conflicting opinions about the necessity of the design attributes—twin engines or two seats—but the Phantom grew to be an extremely successful combat airplane. Within two years, the F-4 and the night Crusader began to compete for the Navy's attention.

The F-4 became the fastest airplane (Mach 2.5) in the world before it deployed to sea. Pilots beat the Skyray's records, setting world speed and time-to-climb records practically every time they strapped into an F-4's cockpit. In the radar there were circuits to allow interpretation of the radar picture out to its sixty-mile range even if jammed by enemy countermeasures. There was an automatic acquisition or lock-on provision for Sparrow tracking, but no track-while-scan circuitry; therefore, only one radar-guided missile could be in the air at a time. Despite its sophistication, the F-4's radar could not distinguish targets below it hidden in ground clutter, nor could the fire-control system control missiles fired at targets below the interceptor.

The F-4 was the first and only Navy fighter that had no air-to-air guns. It carried four Sidewinders and two to four Sparrows for destroying enemy airplanes. The Navy did not design the F-4 for bombing, but by the end of 1961, the weapons and attack tactics development squadron, VX-5, had gathered bomb ballistic and gunsight data for dropping the new Mk-80-series bombs from the airplane. VX-5 also checked the airplane for compatibility with a new series of nuclear weapons.

The first F-4B went to VF-121 at NAS Miramar, California. By December 1960, the squadron started F-4 fleet introduction training for pilots and RIOs flying the interceptor. For East Coast training, VF-101 at NAS Oceana, Virginia, formed Detachment A with F3Ds and about a dozen pilots and RIOs. The Navy selected Commander O'Rourke as the officer-in-charge of that unit. The detachment moved temporarily to NAS Miramar for training with VF-121, then returned to Oceana where it also received new F-4s. The unit soon began instruction of F-4 pilots and RIOs on the East Coast.

The first F-4 squadrons were the Atlantic Fleet's VF-74 and the Pacific Fleet's VF-114. VF-74 deployed to the Mediterranean in the *Forrestal* in August 1962, while VF-114 deployed to the Western Pacific in the *Kitty Hawk* (CVA-63) later that same year.

The skipper of VF-74 was Cdr. Julian Lake, a strong, intelligent leader and electronics expert. Through his tenacity, the squadron received the special logistics support that is often needed when new airplanes enter fleet service. Lt. John Mitchell and other former Banshee pilots joined the squadron. The squadron aircrews and maintenance men trained in VF-121 and later in VF-101 Detachment A. After receiving basic training in the F3D, VF-74's new pilots and RIOs actually trained themselves. The VF-121 and VF-101 Detachment A staffs were busy establishing a standard syllabus for the large number of student pilots and RIOs to follow.

A great part of the training as well as the operational success of the first squadrons was due to the work of the former enlisted aircrewmen who became the first F-4 RIOs. From preparation of lectures, ultimately developing a training syllabus for VF-101 and the fleet F-4 squadrons, to understanding the full potential of the radar system while being jammed or in the worst foul weather conditions, these men proved priceless. Some of the men in the initial group of F-4 RIOs were Ace Webster, Carl Martin, Duke Deesch, and Roger Ferguson.[1]

The F-4 was the largest carrier fighter built up to that time. Because of its higher takeoff and landing speeds, the F-4 required the most powerful catapults and strongest arresting gear in the carriers. Capt. Jerry Miller commanded the *Franklin D. Roosevelt* when the F-4s were being introduced. His ship was a test bed to determine if the *Midway*-class carriers had the proper equipment to handle the F-4 safely. With him leading the effort to exploit the new night fighter, the *Roosevelt* showed that the Phantom could operate from mid-sized carriers.[2] The night fighter eventually pushed the day fighters as well as the Skyray and Demon squadrons off the bigger decks. Within three years, by the beginning of the

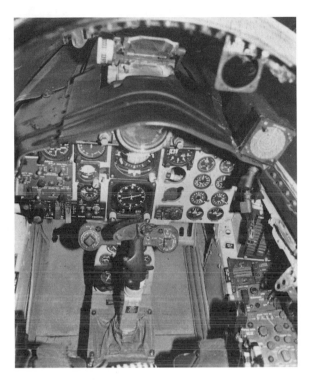

An F-4 pilot's cockpit.
Note the small radar
scope. *National Air and
Space Museum*

Vietnam War, there were two F-4 squadrons on every ship that could
operate Phantoms.

. . .

Six months after the introduction of the F-4, the night Crusader, the
F-8D, finished flight tests and deployed with VF-32 and VF-111 as a lim-
ited-capability carrier night fighter. The F-8D had no search radar but
carried the APQ-83 angle-track radar and one of the first infrared aircraft
detection systems, the AAS-15 infrared scanner. There were hard points
for bomb carriage, but the fleet seldom used the airplane as a bomber
until the Vietnam War.

The F-8E was the next version of the night Crusader. The F-8 series
continued through a "J" model before the airplane retired. The F-8E
solved one big problem. Because the F-4 was heavy and had a high landing speed, the Navy decided not to operate the airplane regularly on

Essex (27C)-class carriers. The night Crusader became the small carrier's night fighter. Ten squadrons deployed with the night F-8 in 1962, and by 1964 the five *Essex* (27C)-class carriers had at least one night Crusader squadron. There were no more Skyray and Demon night fighters in the fleet by that time.

Early Crusaders were unstable in the landing configuration, making carrier landings difficult. Moreover, the night fighter versions were no better. Experienced Crusader pilots said, if the airplane seems to be at the proper landing speed and on glide slope, change something, because the airplane will be falling off the correct speed or the glide slope any-way. Crusader LSOs warned nuggets to search up if they lost the ball, because they would likely be too low. Because of its bad flying qualities in the carrier landing pattern, the decision to use the night Crusader had its deficit side. The F-8 squadrons' accident rate while night flying from carriers was atrocious, 2.7 times the rate of the rest of carrier aviation.[3]

An approach power compensator (APC) was placed in the F-8s to assist the pilots in maintaining proper speed on the glide path. It was partially successful. The air officers in the small carriers contributed to the night Crusaders' problems because in the last years of the F-8's use, the glide slope in the small carriers was set at a higher angle than that used by the large carriers for no apparent reason. That made it even tougher for the night pilot to stay on the glide slope and still keep the Crusader's speed under control.

.   .   .

In the early 1960s, the night fighter squadrons' predeployment training consisted mostly of intercept practice. F-4 aircrews considered flying on instruments rather than by reference to the sea and sky horizon to be nor-mal.[4] Precision flight guided by instruments and radar replaced the more rapid and instinctive moves of fighter pilots who can see the enemy and the horizon. Intercepts were controlled by either air intercept controllers (AICs) on the ground or the aircrews themselves. NAS Oceana and NAS Miramar, the principal fighter bases, were close to controller training cen-ters and ships' operating areas, giving many opportunities for coordi-nation between the night fighters and AICs. Night intercept practice con-sumed about 30 percent of the intercept training time in the air.

Although they were night fighting units, the F-4 and F-8 squadrons did most of their fighter training while at sea during daylight hours. One standard task force exercise prior to the Vietnam War simulated the force's air defense against Soviet bombers. The operation pitted a carrier's fighter squadrons against another carrier's air group or perhaps the Air Force or even another country's air force. The "enemy" flew simulated bombing attacks against the "friendly" task force. Ninety-five percent of the time, the bombers scheduled their time-on-target during daylight hours, attempting to penetrate the task force's ring of CAP stations manned by both day and night fighters.

If the bombers were at high altitude, the carrier controllers and the fighters had an excellent chance of performing effective intercepts on the bombers as the F-4s tried for a Sparrow missile shot at the enemy bombers from beyond visual range (BVR) or the F-8s attempted to close for a rear-hemisphere Sidewinder shot. The great deficiency in the fighters' radars showed up in these exercises. Even the latest Phantom radar could not effectively detect airplanes while looking down toward the ground. All an incoming bomber had to do was fly at low altitude to ensure escape from the carrier's air search radars and the fighter's radar. Later model AEW aircraft had radars designed to detect low flyers. An effective tactic for countering low-flying bombers then placed the fighters at an equal altitude with the bombers in position to detect and get a missile-firing opportunity at the threatening aircraft.

· · ·

By 1960, ten heavy attack all-weather squadrons on the East and West Coasts flew the A-3. One West Coast squadron, VAH-4, supplied detachments of three airplanes each to the *Essex* (27C)-class carriers. The A-3 crews were familiar now with the night launches, night nuclear weapons delivery missions, and night recoveries that the fleet commanders asked them to perform from even the smallest carriers.

Surface-to-air missiles (SAMs) were in the Soviet air defense system by then. Consequently, nuclear weapons delivery tactics used by the A-3 crews had changed. Rather than penetrate Soviet airspace at high altitude and release their weapons in horizontal flight, the A-3 crews now flew at low altitudes and performed a minimum-altitude wingover maneuver similar to the wingover weapon delivery night attack crews had used in the AD-5N. The A-3 crews had only their radars to improve their dead-reckoning navigation on their long-range night missions in any weather. In the 1960s, the weapon system of the best all-weather attack airplane—the A-3—was not much better than those in the obsolete AD-5Ns and AJs or the A-4C, the latest A-4 model.

During the 1960s, the commander of the Southern Region (CinC-South), the major NATO command in southern Europe, held an annual exercise that realistically simulated execution of the NATO nuclear strike plan established to deter Soviet aggression. The CinCSouth staff furnished umpires who observed and worked with the TF-60 staff, providing intelligence about the enemy, bomb damage assessment (BDA), and losses to NATO strike aircraft. Midway through the exercise, the umpire's aircraft loss reports showed that the simulated attrition to the heavy attack squadrons was more than 30 percent. It was obvious that the war planners knowledgeable about Soviet air defenses expected the VAH all-weather units to be at high risk if asked to execute the nuclear strike plans.

During that exercise in 1962, for five days the attack squadrons in the *Forrestal* and *Franklin D. Roosevelt* flew simulated strikes against postulated enemy targets. VAH-1, flying A-3s off the *Roosevelt*, reported that during one sixty-hour period, their aircrews flew over one hundred strike missions, half at night.[5]

The Mediterranean Sea is famous for its "milk bowl" nights. Often there are no clouds; instead, a thick, milky haze reduces visibility. There is no visible horizon, causing pilots to fly most day missions and all night missions on instruments. Consequently, carrier landings at night are difficult. One night in the late fall of 1962, a former fighter pilot assigned to VAH-5 in the *Forrestal* during its 1962/63 Mediterranean deployment and his crew were flying a practice night navigation flight. Although an experienced carrier aviator, he had had little night flying experience. Returning to the *Forrestal* that night in the Mediterranean's milk bowl, he could not get his A-3 aboard. After five attempts, he went to land at the scheduled emergency field, a small Italian strip on the northwest coast of Sardinia.

The strip was on a bluff about 1,000 feet above the sea. There was a low overcast over the field, but there was room below the clouds to land. The *Forrestal* was within fifty miles of the field, allowing the watchstanders in Air Plot to hear the pilot's radio transmissions. The Italians did not answer his calls for landing instructions. Moreover, there were no lights on the strip, although the field had been scheduled for weeks as the carriers' divert field for such emergencies. There was no response when he asked them to turn on the runway lights! His final radio call told the ship that he was making a pass using his radar to line up with the runway. The A-3 ran out of fuel and crashed into the bluff, killing the pilot and his two crewmen. Weather and the night claimed three more naval air warriors.[6]

In 1961, Heavy Attack Wing One received a new, supersonic, two-crew heavy attack airplane—the A-5A Vigilante, an airplane designed solely to deliver nuclear weapons. Its radar and weapon control system were state of the art. Probably the most important innovation of the A-5A was that it was the first Navy airplane with an inertial navigation system (INS). Ballistic missile submarines had used an INS for several years, and by the time the A-5A arrived, some *Forrestal-* and *Kitty Hawk*-class carriers had incorporated a version of the submarine's INS. Data fed from the carrier's INS initiated the A-5A's system, which greatly improved the navigation and weapon delivery accuracy of this all-weather airplane.

The Vigilante replaced the A-3 in some squadrons immediately, going to sea first with VAH-3 in the *Enterprise* (CVAN-65, now CVN-65) during her fleet deployment that began in July 1962. "Heavy Three" participated in a NATO strike exercise into Spain on that cruise. During the lifetime of the airplane, the Atlantic Fleet was home to three A-5A heavy attack squadrons: VAH-1, VAH-3, and VAH-7. The Pacific Fleet did not receive any A-5As. The A-5A was not an effective bomber, remaining in that role for only about two years. At that time the Navy modified the supersonic A-5's weapon system, changed the airplane's mission to reconnaissance, and redesignated it the RA-5.

The advent of the angled deck, mirror and lens, flight deck floodlights, and better radars improved the light attack squadrons' night and all-weather effectiveness along with that of the fighters and heavy bombers. By the early 1960s the AD squadrons had added the night attack mission held previously by VC-35 and VC-33 to their nuclear weapon and day attack missions. The AD squadrons practiced the night laydown and low-angle dive bombing techniques with conventional bombs and rockets perfected in the VC night combat squadrons. Not having a night tactical doctrine at the time, the AD squadrons adapted the VCs' tactics for night interdiction missions. Squadrons still used flares for illumination, although there had been no improvements in that piece of equipment since the Korean War.[7]

The A-4 and A-3 pilots and crewmen had started practicing a new night and all-weather nuclear weapons delivery tactic in the late 1950s. The tactic, called "buddy bombing," was designed to enable an air group to get more nuclear weapons into hostile territory if the United States went to war. Rear Adm. John Hyland at the Joint Strategic Target Planning System headquarters wanted an increased Naval Aviation presence in the nation's nuclear weapon strike plans. When he heard that A-4s had tested flying on the wing of a radar-equipped A-3 deep into an enemy's territory at night and in any weather condition, he asked for proof of the buddy-bombing tactic before scheduling such missions in the nuclear weapon strike plans. One of the first A-4 squadrons, VA-44, demonstrated the tactic to Hyland and the joint targeting group in 1958. The staff was convinced and incorporated the tactic into the U.S. nuclear strike plans.[8]

Buddy bombing with an A-3 on nuclear weapons delivery missions was one of the toughest night and all-weather tasks for A-4 pilots. The A-4 pilots needed to fly with Blue Angel precision in the worst conditions to carry out those missions, a task more difficult than the pathfinder missions that the night ADs and AD bombers had flown together earlier in the decade. The task was more difficult because the AD team flew the same type of airplane, making formation flying easier.

A typical buddy-bombing flight began with a radar-equipped, multiseat A-3 acting as pathfinder for a single-seat A-4 flying tight formation on A-3's wing, day or night in any weather condition. After a low-altitude—usually 500 feet above the terrain—penetration of enemy airspace, the A-3 would use its radar to locate the A-4's target. Dropping to 200 feet above the terrain and maintaining 500 knots, the pathfinder would turn to the A-4's weapon delivery heading. At the proper distance from the target for the A-4 to perform an accurate nuclear weapon delivery maneuver, the A-3 pilot would flick his lights on, signaling the A-4 to detach from his wing.

Instantaneously, the A-4 pilot had to switch his attention from the A-3's dim formation lights, which had been slightly above and to the side of the A-4's cockpit, to his own instruments. He had to immediately scan his attitude indicator, altimeter, and directional gyro to ensure that he maintained altitude, airspeed, and heading. Separating from the A-3 demanded the highest degree of skill from the A-4 pilots to maintain altitude while switching to instruments as the A-3 turned away toward its own target.[9] Most pilots gained at least 500 feet in the process. The A-4 pilot had to maintain heading and settle on the proper height above ground to complete his half-Cuban-eight weapon delivery maneuver, ignoring darkness, bad weather, or, if at war, gunfire.

The early A-4As and A-4Bs had poor attitude indicators, barely capable of guiding the pilot through such a maneuver. There was another instrument, part of the loft bombing system, designed specifically to give more precision to nuclear bombing. That gyro instrument, when engaged, showed the pilot the proper amount of g's to pull into the loop and his airplane's roll attitude. It was much more reliable than the attitude indicator used for normal instrument flying in the early A-4s. Moreover, the A-4 had extra power to carry the airplane over the top of the loop, giving the pilot plenty of room to roll upright on the back side of the half Cuban-eight.

During the years that the A-4s and A-3s practiced buddy bombing, no casualties occurred during night training missions. Experienced pilots have differing memories of the experience of buddy bombing, probably because of a difference in their night flying experience before being introduced to the tactic. One pilot who had been in VC-33 flying night ADs before transferring to A-4s, flew buddy-bombing missions with the A-3s of VAH-6. In 1962 VAH-6 was one of the first Pacific Fleet A-3 squadrons to team with A-4s. He said that the worst part of buddy bombing was not separating from the A-3 nor the night delivery maneuver, but rather was the task of convincing the A-3 pilots to reduce their speed so that the A-4 could keep on its maximum-range speed schedule.[10]

•   •   •

The third model of the Scooter, the A-4C, was introduced in 1960. Designed to be a night-capable airplane, the changes from the earlier A-4 models included an improved attitude indicator, a radar altimeter, and a ground mapping radar to assist in navigation. The radar also had a primitive ground clearance, terrain avoidance mode. That radar mode, plus the radar altimeter, enabled the pilot to estimate his altitude over the ground and avoid crashing. Fleet commanders hoped that with the new mapping and terrain avoidance radar, the A-4's single pilot would be nearly as effective as the old multicrew AD-5Ns and heavy attack airplanes.

During the summer and fall of 1960, VX-5 conducted a test project to determine if the single-seat A-4C could deliver nuclear weapons alone in instrument and night conditions.[11] The first project task was to find out if the A-4C's radar could distinguish between different ground features, natural and manmade. The three pilots assigned to the project flew enough sorties over the mountains and valleys of California to answer two questions: was the radar good enough to help the pilot navigate over land at night and in bad weather? and was the radar satisfactory for acquiring and targeting major land facilities such as airfields and industrial complexes? The answer to both questions was yes.

Independent of the VX-5 project, Lt. Cdr. Bob Mandeville, the operations officer of VA-83, began training squadron pilots to use the new A-4C's radar. They practiced identification of their planned landfall locations on simulated nuclear weapon strike missions. Mandeville believed, like VX-5, that an experienced night fighter or attack pilot could interpret the radar picture and locate large manmade facilities such as airfields or large factories.

The next task in the VX-5 project was to determine if an average pilot could fly at low altitude over land while operating the radar in instrument conditions or at night. Each of the three project pilots flew fifteen hooded sorties chased by a safety observer over low-altitude routes in the desert and the foothills of the Sierras. This type of flying was similar to the VC-35 night strike practice missions, but flown at more than twice the speed. The test team pilots found that, with practice, pilots could be safe flying on instruments, but they were uncertain that a pilot in unfamiliar territory could fly the A-4C safely alone during bad weather.

The techniques of radar navigation were used during a 1961 NATO strike exercise by Cdr. Jack Herman's VA-66. Lts. Rupe Owens and Jerry Breast and other pilots used the terrain avoidance mode of the A-4C's radar during foul weather conditions to complete their simulated nuclear strike missions. Reports from VA-66 strongly supported the use of sensors and electronics to meet the challenge of night and all-weather strike operations.[12]

The VX-5 avionics and weapon system technicians designed a new radar circuit for the A-4C's radar similar to the one Lieutenant Shaw had placed in the APS-31 that enabled the radar operator to measure the distance to a significant radar target. Knowing the distance to his target, the pilot could deliver his weapon using the loft bombing maneuver. In the last series of A-4C radar project flights, the VX-5 test team evaluated the instrument-flight loft bombing technique on a practice bombing range. The project team flew those sorties either at night or using an instrument hood. Flying hooded was still just as difficult as when Lieutenant Akers first used a hood in the 1920s.

The run-in to the target began about ten miles from the target bull's-eye. After locating the radar image of the target on the scope, the pilots went under the hood, descended to 500 feet above the terrain, increased speed to 500 knots, and proceeded with the delivery maneuver. The test team found that an experienced pilot could execute the loft bombing maneuver and achieve acceptable accuracy with the bomb while flying on instruments. VX-5 was actually testing the feasibility of pilots using the same loft bombing maneuver that the fleet A-4 pilots had been performing for a few years while buddy bombing at night. In this test, however, performing the half Cuban-eight actually required less concentration because the pilot did not have to switch his visual scan from an A-3 to his instruments just before attempting the loft bombing maneuver.

On one flight while under the hood in the delivery maneuver, the test pilot's attitude indicator—his principal reference to the outside world—precessed. While he guided his A-4 by false indications, off heading and in a bank, the practice bomb released, arcing about a mile and a half! The pilot finished the maneuver safely, although badly off heading, not knowing about the miss until the tower figured out what had happened and reported his bombing error. This flight clearly showed that for effective bomb deliveries using instruments in night or any weather condition, there can be no system malfunctions.[13]

The work done in VX-5 had focused on testing the A-4C system's ability to find and attack targets on land. There was no doubt that the A-4C and later A-4 models could easily find and put a bomb into a ship target at night and in all weather conditions. However, just as in World War II, identification of an individual ship with the aircraft's radar at night or in adverse weather was not yet possible. The final VX-5 project report concluded that although the A-4C had an acceptable radar and good instrumentation, the aircraft demanded too much skill from an average fleet pilot to make it an effective night and all-weather nuclear weapons delivery airplane without a pathfinder.

The Navy's aviation weapon test facilities at China Lake, California, acted as an incubator for inventors in the 1950s. Among the new ordnance equipment conceived there was the multiple bomb rack (MBR), which allowed small jet fighter-bombers such as the A-4 to carry up to eighteen 500-pound bombs. Without the MBR, the A-4 could carry only three bombs.

The MBR laid around VX-5 as a curiosity until John F. Kennedy became president in 1960. He renewed interest in the use of conventional weapons, an interest that started Naval Aviation on conventional weapons' training programs. The Navy had completely changed its inventory of combat aircraft and had redesigned the conventional, general-purpose bomb since the Korean War. The MBR promised to give the new fighters and attack airplanes much greater effectiveness. Now, Naval Aviation needed new tactics for the entirely new airplanes and bombs.

VX-5 was told to take charge of developing new tactics for delivering conventional weapons as well as nuclear weapons. One of the principal project tasks was to develop and codify night tactics for the A-4. The A-4 squadrons up to that time had done little-to-no night work with conventional weapons. The test team modified the night dive bombing techniques, with and without flares, and night interdiction tactics that had been standard in VC-35 to accommodate jet airplane performance. In a jet, the pilot had to work harder because the airplane was significantly faster than the propeller-driven AD. Flying flare-equipped A-4s, the project pilots also tested night close air support procedures working with Marine Corps FACs at Camp Pendleton, California.[14]

. . .

In 1963, the commander of Carrier Division Four, Rear Admiral Hyland, had the task of conducting ORIs of the Atlantic Fleet carriers before the ships deployed to the Sixth Fleet in the Mediterranean. He was aware that the A-4 squadrons might not be getting the training they needed to effectively carry out the night attack mission that was falling on their shoulders as the ADs left the fleet. It was suggested that it was impossible for all the pilots in an A-4 squadron to get enough night experience— even on the bigger, angled deck carriers—to be effective in combat, and therefore night teams should routinely be organized in each squadron, just as night teams operated in fighter squadrons near the end of World War II. Hyland considered this recommendation but did not forward it

140

to higher authority. Hyland foresaw the time when *all* carrier aviators would be fully qualified for night air combat and felt that the creation of night teams would delay that goal.

Hyland did support the idea of adding night interdiction flights, proven in the Korean War and using the tactics developed in VX-5, to the A-4 squadrons' ORI flight schedule. Soon thereafter, Commander, Naval Air, Atlantic (ComNavAirLant) added a series of night attack training flights to the standard A-4 training requirements. The Pacific Fleet A-4 squadrons also added night bombing and interdiction tactics to their training requirements at the same time.

The A-4 squadrons had gotten a few multiple bomb racks by the summer of 1963 and, on their own initiative, were practicing conventional weapon dive and glide bombing day and night. The biggest change for A-4 squadrons to meet the new ORI night attack requirement was to practice night interdiction tactics. In November 1963, the *Forrestal's* air group, including VA-83, received its ORI. The wingman of a VA-83 night mission section was Lt. Bill Westerman. After a night flight over an unfamiliar, hilly piedmont region of North Carolina following the trace of a country road, his eyes were big with excitement at meeting the challenge of night flying. Westerman later joined the A-6 community and became well known for his ability as a night attack pilot during and after the Vietnam War.[15]

．．．

During the Kennedy administration, the Soviet Union set up several ballistic missile sites and SA-2 SAM sites in Cuba. Although the president negotiated a halt to Russian military activity in Cuba without starting a war, the Navy began rehearsing air strikes that could be executed against Cuba in the event the country did go to war to evict the Soviets from the Western Hemisphere. Replacement Air Group Four, commanded by Capt. Jim Ferris, took responsibility for planning and coordinating that contingency mission.

The captain's strike plan called for one or more air groups to move to NAS Cecil Field, Florida, when the National Command Authority (NCA) issued a preliminary warning order for an air strike against Cuba. East Coast air groups flew several missions simulating a dawn attack on Cuban targets after a low-altitude approach across the Florida Strait. When ordered, twenty-five to forty aircraft would launch in the predawn

141

hours. After joining in relatively close formation to allow visual contact with the leader in the darkness, the strike would proceed toward Cuba. The night "group gropes" were challenging even to the best pilots, requiring a high degree of skill and concentration.[16]

. . .

The process of developing a new Navy airplane often takes many years. To begin that process for a new night attack airplane, Lt. Cdr. Mark Hill, Lt. Cdr. Harry O'Conner, and Lt. Bill Offtermatt, who were assigned to VC-35 in the middle 1950s, put all the accumulated knowledge from World War II and the Korean War about night attack aircraft needs into a paper for BuAer. Their ideas became the core of the 1957 specification for the first jet-powered night and all-weather attack airplane.

Surprising opposition to a jet for night attack came from Capt. Bill Martin. He and many others believed that the Navy should continue operating attack airplanes with reciprocating or turboprop engines.[17] Martin was correct in one respect. Locating targets at night, particularly enemy vehicles or troop targets, can be best done at the slower speeds of propeller-driven airplanes. However, the Navy learned during the Vietnam War that the slower ADs could not survive modern antiaircraft defenses. The Navy made a sound decision to proceed with a jet as its next night attack airplane.

Grumman won the contract to design and build an airplane to meet the BuAer specification for a new night and all-weather attack jet. In February 1963, six years after the birth of the concept, that company delivered the first twin-engine, two-seat A-6A Intruder to VA-42, the newly designated medium attack Fleet Replacement Squadron (FRS). It was a remarkably fast airplane, reaching a top speed of about 500 knots with almost its full load (14,000 pounds) of bombs aboard.

The A-6 had its two crewmen seated side-by-side in a huge cockpit. The pilot's instrumentation included the first cathode ray tube display, a vertical display indicator (VDI). The glass screen display showed the airplane's attitude and relative speed in reference to representations of the sky and ground. The nickname for the A-6's center display was "pathway in the sky." A more conventional instrument, the horizontal situation indicator (HSI), presented information from the radio navigation system. Use of thermometer-type gauges for engine instruments was another innovation. Old-style instruments backed up the radically new primary flight instrumentation.

The B/N operated the radar, weapon systems, and ECM equipment. The first computer-controlled weapon system, designated the Digital Integrated Attack Navigation System (DIANE), was the brain of the A-6A. DIANE, like the navigation system in the A-5, functioned with an INS at its heart; therefore, only carriers fitted with an INS could operate the A-6 to its full effectiveness. The radar picture viewed by the B/N was a PPI for navigation and target acquisition or E-scope presentation for terrain clearance and avoidance. A Doppler radar allowed the navigation and weapon delivery systems to make wind corrections. The crew could select several modes of weapon delivery: straight-path at high or low altitude, dive, or toss. Once committed to a bomb run by the pilot and B/N, the A-6 system would release the selected weapons when the airplane was in the correct position at the correct attitude and speed for a hit on the target's center.

The Intruder's flying qualities were superb. Carrier landings, even in the worst weather or night conditions, were relatively easy. Rear Admiral Hyland, still Commander, Carrier Division Four, asked for and received permission to fly the new A-6 during the first month it was in the fleet. After his first flight, Hyland declared, "It's a grandmother's airplane!"

The A-6 quickly replaced the A-3 and the A-5A as the Navy's principal night attack and all-weather airplane. The Navy chose new A-6 pilots from the light attack and fighter communities, curiously bypassing the former VC-35 and VC-33 night attack pilots. Those experienced night attack pilots became leaders in night light attack operations. The group chosen for A-6s included experienced carrier jet pilots such as Jack Herman, Don Ross, Bob Mandeville, Rupe Owens, "Buzz" Ellison, "Sweet Pea" Allen, and Bill Westerman, all of whom had jet backgrounds before entering the world of dedicated night and all-weather attack flying.

The first skipper of an A-6 training squadron was Commander Herman, who took the job when VA-42 transitioned from training AD pilots and maintenance personnel to training A-6 crews. Ross was directly responsible for setting up and getting the A-6 training syllabus started. Mandeville became the training squadron's operations officer shortly after the squadron began flying. The squadron received its initial cadre of bombardier/navigators from the A-3 squadrons. Roger Smith, Ted Been, and other experienced A-3 B/Ns, like their RIO contemporaries in the newly forming F-4 squadrons, passed on their knowledge of radar and weapon systems to the new night and all-weather pilots.

The NATC course for prospective Naval Flight Officers (NFOs) looking forward to carrier all-weather units trained the students to be B/Ns in the A-3s, A-5s, and now A-6s, as well as training F-4 RIOs. After initial ground school indoctrination, all student NFOs flew twelve flights with an instructor pilot in the T-2 Buckeye jet. These flights separated the students who could not endure maneuvering flight from those who could be assigned to carrier aircraft. The men who could stand the stress of flying in combat aircraft began training to fill carrier airplane seats. The prospective NFOs then learned dead-reckoning, radio, and radar navigation techniques, followed by lessons in basic radar techniques applicable to either all-weather fighter, medium attack, or ASW missions. When selected for the new medium attack squadrons, the nugget B/Ns joined VA-42.

When the squadron was first formed, the VA-42 aircrews flew training missions in two A-6s that had been stripped of their radar and weapons systems. These airplanes had been used previously for airframe testing at NATC. Instructor pilots and B/Ns practiced normal flight procedures and some "iron sight" dive bombing until they received the first full system-equipped airplanes in June 1963. The squadron did not even have an A-6 weapon system simulator for a year, but a Grumman representative, Don King, helped to build a cardboard facsimile of both sides of the A-6 cockpit. Training in that primitive simulator at least gave the pilots and B/Ns a chance to get familiar with the procedures, dials, buttons, and switches in the airplane.

Student A-6 pilots and B/Ns had to be trained by instructor B/Ns and pilots, respectively, familiar and experienced with the functions performed on both sides of the Intruder's cockpit. King, a former A-4 pilot, had become an A-6 weapon systems expert, equally at home in the pilot's or B/N's seat. He started the cycle by training first the VA-42 instructors until all twelve instructor pilots and B/Ns were fully capable of training students.

In the meantime these men were writing lesson plans for ground school and flights for the new A-6 syllabus. By the fall of 1963, the train-

ing squadron had an A-6 Weapon System Trainer, a relatively sophisticated flight simulator. They flew Gulfstreams, TC-4Cs, for B/N training.
The Gulfstreams were configured to train several B/Ns in the cabin part
of the airplane. The TC-4C was a successor to the SNBs used to train the
first night fighter and TBM radar operators in World War II. By the time
the Vietnam War began, VA-42 had a proven training syllabus.[18] In 1968,
a second A-6 FRS, VA-128, was finally established at NAS Whidbey
Island, augmenting the training capacity of VA-42.

The philosophy in VA-42 was that a pilot and B/N should fly together
at all times. The fleet squadrons tried to keep the pilot and B/N relation-
ship permanent as much as possible, but sometimes sickness, injuries,
or scheduling difficulties broke the crew's partnership temporarily. Cdr.
Jerry Zacharias entered A-6 training in VA-42 as a student and prospective squadron commanding officer during 1965. He had never flown
with a crewman and had little tactical experience in night air combat
operations. He felt that VA-42's excellent ground and flight training
courses eased the introduction to night—or as now designated, medium—attack.[19]

Chapter 7 **THE VIETNAM WAR**

By the time the Vietnam War started, most carrier pilots and airplanes were fully capable and qualified for operations at night and in foul weather conditions. A few fighter squadrons still flying the day Crusader deployed in the five *Essex* (27C)-class carrier air wings (as the air groups were renamed in 1963), but each of those air wings had at least one night Crusader squadron. At least two A-4 squadrons and an AD squadron were aboard every attack carrier. F-4s had replaced the Demons and Skyrays in the bigger carriers. A full A-3 bomber squadron deployed in each big carrier and A-3 detachments deployed on selected *Essex* (27C)-class carriers. Although the fighter and heavy attack squadrons still prepared only for operations in a nuclear war with the Soviet Union, the light attack squadrons spent about 40 percent of their training time practicing day and night conventional weapons delivery techniques. The carrier Navy was prepared for night and all-weather operations in the next war.

That war began in earnest on the afternoon of 2 August 1964 when North Vietnamese torpedo boats attacked the U.S. destroyer *Maddox* in the Tonkin Gulf. Tensions had been rising between the United States and Ho Chi Minh's Vietnamese Communists since the end of World War II. The attack on the U.S. ships opened up the festering trouble with the North Vietnamese. Three days later the *Constellation* (CVA-64) and *Ticonderoga* launched air attacks against North Vietnam's torpedo boat bases. The war thus begun in Southeast Asia to punish the Vietnamese Communists continued sporadically for almost nine years.

Besides providing ground troops and other support to aid the South

146

Vietnamese army, the United States began the Rolling Thunder air campaign against the North Vietnamese in March 1965. The NCA in Washington, D.C., directed the campaign, selecting the targets, the times-on-target, and sometimes the preferred ordnance for the Rolling Thunder "Alpha" strikes. The carrier commanders had little flexibility in using naval air power. The war planners preferred that the air wings execute large Alpha strikes rather than strike in small groups because in their opinion the Alpha strikes promised more damage to the major targets than any other tactic.

An Alpha strike group included between eight and sixteen attack airplanes and as many as twenty-four fighter and support airplanes. However, a successful strike had to have good weather to locate and bomb the target. On good weather days during Rolling Thunder, each carrier air wing usually flew up to three Alpha strikes.

The North Vietnamese opposed the Rolling Thunder strikes with a few Russian-manufactured MiGs, but until the Russians gave SA-2 SAMs to the North Vietnamese in April 1965, antiaircraft guns were the principal air defense weapon with which combat pilots had to contend. After the SAMs came in, the air wings stopped high-angle dive bombing, but did not change their night attack low-angle and laydown tactics, except to ensure that the aircrews could be constantly aware visually of the presence of airborne SAMs. Rarely did the airplanes fly over a broken or solid cloud deck, day or night. New ECM devices, together with evasion maneuvers started upon visual detection of airborne missiles, forced most SAMs to miss and kept aircraft losses from the SAM threat very low.

The Southeast Asia monsoon cycle that would heavily influence U.S. air activity had its first effects in the fall of 1965. Annually, beginning in late October, the North Vietnamese weather turned bad. The rainy winter monsoon lasted until late March or early April, although some rare breaks appeared in the poor flying weather. The seasonal weather was characterized by low ceilings and rain that prevented effective visual air strikes. It was difficult to get even a section of light attack aircraft into North Vietnam during the winter without exposing them to high risks from enemy ground-to-air defenses. The weather at the ship was not as important, because the carriers and air wings had perfected departure and approach procedures that permitted operation of large strike groups in bad weather.

The NCA scheduled few Alpha strikes during the winter monsoon. If good weather had been predicted but it turned bad, Washington delayed authorized Alpha strikes, giving the commander of Task Force 77 authority to set a new time-on-target. Sometimes, it would take CTF-77 several days to launch a delayed Alpha strike.

There were many occasions, particularly during the winter monsoon, when TF-77 commanders launched Alpha strikes, letting the strike leaders make the decision of whether or not to lead their strike into North Vietnam in foul weather conditions that greatly favored the enemy. I was executive officer of an A-4 squadron, VA-112, in the *Kitty Hawk* when she arrived in the Tonkin Gulf near the beginning of the 1966/67 winter monsoon. During that and a second combat deployment, I led several Alpha strikes in marginal weather conditions. I will describe one strike typical of that type of operation.

. . .

The target was the Haiphong thermal power plant. The strike group included twelve light and medium attack bombers loaded with 1,000-pound bombs, eight other A-4s as flak suppressors and SAM suppressors carrying Shrike antiradiation missiles (ARMs), and four F-4 fighter escorts. An E-2 Hawkeye AEW aircraft was airborne to relay information between the strike group and the *Kitty Hawk*. KA-3 tankers were airborne for refueling any aircraft, if necessary. (By that time, the A-3s had transferred their bombing mission to the A-6s.)

The ship launched a weather reconnaissance airplane about one and a half hours before the strike's scheduled launch time. At launch time, the weather airplane reported several cloud decks extending from the middle of the Gulf of Tonkin inland as far as he could see. The clouds definitely covered Haiphong. The bottoms of the clouds were at about 2,000 feet, with the tops at about 20,000 feet. Believing the target weather would improve, the *Kitty Hawk's* captain ordered the strike to launch. The E-2 was off the deck first, followed by the fighters, the A-4s, and the A-6s. The KA-3s launched last.

My A-4 was about the tenth airplane launched. I pulled the gear and flaps up, called departure, and disappeared into the clouds. Scanning across the altimeter, airspeed, turn-and-bank, attitude, and rate of climb indicators provided a comfortable feeling that all was going well on the climb-out. Departing airplanes had enough interval between them to

148

VA-112 A-4s ready for launch from the *Kitty Hawk* during the Vietnam War.
U.S. Navy

prevent midair collisions while climbing through the clouds. I had set the strike group rendezvous to the north of the ship at 20,000 feet, above the reported cloud tops. When I skimmed out of the clouds, I could see airplanes in front of me already joining and setting up the strike formation. The fighters were above the center attack group. The flak and SAM suppressors were on the flanks. Joining on my A-4s, I moved into the lead.

When the A-6 division leader radioed that all his airplanes were formed up, I reported that we were departing for Haiphong and switched to the ship's weather reporting frequency. The weather scout continued to report marginal conditions near the coastline, but it seemed that the cloud decks were breaking. As I continued toward the point at which we would come into SAM range, I switched back to my group's tactical radio frequency to see if there were any problems. All was quiet. Soon I had a good picture of the weather between my strike group and Haiphong. I switched again to the ship's frequency, checking for further instructions, but the ship had no guidance. To proceed or not was my decision. From where I sat, completing the strike was worth the risk. I called the group, saying something like, "We're going in!"

I began a slight descent, dodging thicker clouds and following the landmarks toward our planned dive roll-in point by the power plant. A few SAMs, which looked like flying telephone poles, flew up through the lower cloud deck into the formation, but none hit. Either the electronic countermeasures were working or the Shrikes were forcing the SAMs' guidance radars to shut down after the SAM sites fired. As the flak suppressors turned away to look for flashes of active antiaircraft gun sites, a clump of clouds drifted over the spot where I wanted to start our dive on the power plant. Sliding around the clouds, I rolled in, followed in turn by the other eleven bombers, each varying its dive heading a bit from the airplane ahead. The spread of airplanes in a fan pattern created a more difficult target for the gunners. The antiaircraft fire was heavy, but we were lucky on that strike. Perhaps the gunners had not expected us to get through the breaking overcast. The fighter escort was alert but inactive— no MiGs that day.

After the attack, the group collected itself in sections of two airplanes, with each section leader reporting when overwater to the E-2. The E-2 reported to me and the ship when all the aircraft cleared North Vietnam. One of the F-4 fighter escorts had taken a hit in his starboard wing but was apparently in good enough condition to attempt a carrier landing. One of the tankers joined with the damaged F-4's section in case he needed fuel.

The rest of the strike group returned directly to our marshal. At the appropriate time, the ship's controller started the strike airplanes' approaches. Because the overcast was above 1,000 feet, we let down below the overcast in divisions and entered a visual weather condition landing pattern. Without the all-weather training that the group's aviators had, the strike would not have been possible.

· · ·

When the Rolling Thunder operation began, CTF-77 established Barrier Combat Air Patrol (BARCAP) stations to protect the ships of the task force from possible MiG attacks and from PT boat attacks such as the one that precipitated the war. Even during periods when the carriers

150

withdrew from the Tonkin Gulf to support the ground war in South Vietnam, the BARCAP continued to protect the cruisers and destroyers that operated farther north than the carrier stations. The fighter squadrons tried to fill two to three BARCAP stations twenty-four hours a day, but there were some intervals when no fighters were airborne. Backing up the airplanes on BARCAP stations were aircraft in alert status on the carriers. The BARCAPs saw little air combat, for there was almost no North Vietnamese air threat to the fleet. The MiGs spent their flight time defending their homeland against the Navy and Air Force strikes.

Early in the *Kitty Hawk*'s deployment, on 20 December 1966, I was walking to my airplane for a mission on a night with a low overcast and poor visibility when the flight deck bull horn roared, "Launch the Condition Five CAP!" The huffers hooked to the alert F-4s started the plane's engines. Within a few minutes, VF-114's Lts. Denny Wisely and Dave Jordan, pilot and RIO, respectively, along with VF-213's Lt. Dave McCrea and Ens. Dave Nichols, took off with their F-4s' afterburners lighting up the night. "Red Crown," the cruiser with air raid warning duty, had detected two slow-speed air targets moving south over the Gulf along the North Vietnamese coast about fifty miles south of Haiphong. Red Crown designated the bogies as hostile enemy aircraft, "bandits," because there were no friendly aircraft airborne except the night fighters that had just launched.

Closing rapidly on the bandits from an initial range of about seventy-five miles, both crews armed their AIM-7E Sparrows for a BVR shot. Apparently North Vietnamese radar detected the interceptors because the bandits turned toward the shore—maybe into a SAM trap for the F-4s. Before the enemy reached the shoreline, Jordan detected and locked on a bandit at ten miles, and Wisely fired. McCrea saw the flash of the Sparrow launch and watched the bandit explode about five miles from Wisely's aircraft. McCrea's RIO, Nichols, locked on the other bandit. Seconds after the first kill, McCrea fired and knocked down the second enemy aircraft.

Post-action intelligence reports identified the aircraft as North Vietnamese AN-2 Colts. I made a brief entry about the night fight in my war diary: "Our condition CAP (F-4s) was launched last night and 'Sparrowed' two bogies over water. I listened to the intercept—good job." The action brought the first night successes for carrier-based fighters since

A VF-114 F-4, the night fighter's workhorse, lands on the *Kitty Hawk* during the Vietnam War. *National Air and Space Museum*

World War II. Moreover, the night kills were the first Navy air-to-air victories in BVR conditions.

. . .

When there were no Alpha strikes scheduled during good weather, the light attack and fighter squadrons, if not all on BARCAP, flew interdiction missions into North Vietnam. These missions concentrated on the railroad and road networks south of Hanoi. There was only one Vietnamese railroad feeding the south, and daylight strikes quickly put it out of action. The enemy made little effort to repair the railroad. Effective interdiction of the road network was a different matter. Keeping the roads closed was tough for several reasons. First, the Viet Cong and North Vietnamese troops in South Vietnam used even fewer supplies than did the North Koreans about fifteen years earlier. Moreover, the North Vietnamese jungles allowed the few needed supply trucks to hide quickly if the drivers heard our airplanes.

152

It was only on days with no Alpha strikes but good weather in North Vietnam that there was any significant carrier night activity until the A-6 entered the war. Night interdiction was more difficult than during the Korean War because of the trucks' ability to hide and because the jets' speed reduced the chances of a night bomber finding dim truck lights. Making night interdiction missions even harder, the North Vietnamese kept a strong antiaircraft gun and SAM defense along the roads and trails.

On 21 January 1967, about a month and a half after my first combat cruise began, I flew my first night interdiction mission into North Vietnam. There had been no Alpha strikes scheduled during the day, but the winter monsoon's overcast rose to about 15,000 feet late in the afternoon. My section's mission was to locate and destroy trucks or truck parks along North Vietnam's Route One, the major road connecting North and South Vietnam, located a few miles inland from the coast. I briefed my wingman in the red light of the ready room. Squadrons still briefed under red ready room lights at night, although aircrews lost most of their night vision under the flight deck's moon lights.

Coming out of the island hatch, the floodlights cast shadows around the crowded flight deck, but it was easy to pick out my aircraft. The VA-112 Broncos' airplanes, distinctive with a black stenciled horse's head on the fuselage, were as usual parked tail outboard along the deck edge toward the after part of the flight deck. A thorough preflight check, using my flashlight, of the airplane's exterior and the six 500-pound bombs preceded the seven-foot climb up the ladder to the cockpit. Illuminated by the floodlights, I settled into the cockpit, tightened the seat belt to my harness, and connected the g-suit. I put on my helmet and plugged in the radio cord and oxygen mask. I was ready to go. I had flown several combat missions before I realized that by the time I'd gotten strapped in my airplane, ready for engine start, most of my anxiety about the coming mission was gone. By that time in my career, the night was not so mysterious.

After the engine start and aircraft system checks, I checked in with Prifly and departure control. The E-2 "Hummer," so called because of the loud, humming sound its turboprops made at idle, launched first at every cycle to relieve a squadron mate on the AEW station. The A-4s followed the fighters. The taxi director guided me smartly with his wands onto one of the bow catapults. I ran the engine up to 100 percent, turned on my lights, and trusted the catapult officer. With my head braced against the headrest, elbow against my stomach, and hand resting behind the stick, the A-4 jumped off the cat easily despite a heavy load.

On the gauges going off the catapult, I checked the plane's altitude and attitude, raised the gear and flaps, and proceeded to rendezvous with my wingman, who had gotten off the deck before I did. His lights appeared where I expected them to be, west of the ship about ten miles away below the overcast. I had decided it was not worth the time and trouble punching up and down through the clouds before going "feet dry" over the North. The *Kitty Hawk* was not that far from the coast.

I took the lead with my wingman in a loose trail, keeping track of my airplane's dorsal and dim blue formation lights. When we reached the beach at a point near the middle of our route, we turned out all lights and my wingman began dropping back. I had chosen to let each of us search independently, but as leader I called all route and altitude changes. We had enough airspace below the overcast to see and evade missiles if the enemy fired SAMs at our airplanes. It was clear and dark underneath, but with no horizon. At 1,000 feet above the terrain, I could barely make out the winding path of Route One through the trees below. I turned north.

We bypassed Thanh Hoa, patrolling north toward Ninh Binh. If either my wingman or I located a promising target, the one finding the enemy would drop flares and call for the other to join the attack. I could see bursts of rapid-fire 23-millimeter tracers around the Ninh Binh bridges, but nothing close to us. Those 23-millimeters at night really attracted my attention. That gun's rate of fire was so fast that the stream of bullets, always with tracers at night, looked like burning water from a fire hose. At the end of the stream's trajectory, it sort of petered out, tracers dropping like a fancy fireworks burst.

I reached the northern end of our route and swung west. When my wingman called his turn at the same spot, I headed back to Route One and started south. Neither of us had seen anything. Near where we had started our interdiction mission, I saw flickers of light. There were moving trucks coming toward me, probably empty, but at least moving. Checking that I had armed my bombs and 20-millimeter cannon, I pulled back to idle power and started a strafing run on the lights. The guns jarred the little A-4. I went on the gauges during the pullout from the dive, then climbed, making a tight left turn, and released two flares.

I called my wingman, reporting my luck in discovering the trucks. He answered that he understood and soon reported that he saw my flares and was joining me. I had seen no gunfire. I ordered two passes apiece, telling him to start his bomb run from south to north after I called that I was off the target. My wingman and I would be aiming only at a spot on the road under the flares, for I had seen nothing after my strafing run. After my wingman called rolling in, I waited for him to call off target— and waited. Nothing! He never called.

By that time, the flares had extinguished themselves. I had seen no secondary explosions from our bombs or any other explosion that might have been an impact from an airplane. After dropping my other four flares, I circled just above the light in the hopes of seeing something under them. I climbed and reported my wingman's disappearance and possible loss to the airborne E-2 controller. He sent a tanker toward my location and vectored an F-4 section to help search. In the darkness, I had them take altitude separation on me and we flew over the area for about thirty minutes. We saw no fires on the ground and heard no emergency radio transmission.

I called the KA-3 tanker, letting him know that we needed fuel. As the F-4s and I approached the refueling position, I saw the tanker's bright and flashing lights about ten miles away as it proceeded on the refueling course. The F-4s refueled first, so I stayed away from the tanker and watched the lights of the F-4s take position behind the A-3. After the second F-4 finished refueling, I accelerated, slid into position, and took my turn refueling.

Refueling from an airborne tanker's hose at night requires as much precision as any maneuver in an airplane. The refueling hose, called a drogue, drops behind and below the tanker. It has a basket at the end that has the fueling fitting at the joint of the basket and the hose. The pilot of the refueling airplane must put his airplane's fuel probe into the bobbing and weaving basket, making a firm connection with the fitting. At night there are only four lights on the rim of the basket to guide the refueling pilot. The A-4s had been tanking for each other on buddy strike missions for years. Tanking from an A-3 was easier because the larger airplane was a much more stable platform. An A-3 tanker's drogue basket was almost still, except in very rough air. After leaving the tanker and reaching marshal, the approach, CCA, and landing were routine.

That mission was typical of the many light attack night missions flown into North Vietnam with one terrible exception. A pilot and an airplane had been lost. I had probably been looking in another direction when my wingman flew into the ground in his bomb run. Explosions from airplane impacts sometimes do not leave lingering fires. Resuming the search the next day, we saw no evidence of what happened to the airplane or pilot.

• • •

One good weather night sometime during VA-112's two combat cruises, I catapulted into the dark on another interdiction mission. My wingman and I both carried cluster bomb units (CBUs), a fragmentation weapon effective against trucks. This time the track followed the road from Route One toward Hanoi in the valley of the Red River north of Nam Dinh. In the valley, all we had to worry about was SAMs and antiaircraft gunfire, not the mountains bordering the Red River delta.

I led us into North Vietnam south of Ninh Binh and flew west and north about five minutes around that city and Nam Dinh to get to our assigned route. Carrying CBUs that had to be released below 1,000 feet to be effective, I stayed low along our route, hoping that the gunners could not see us. High clouds blocked the moon and starlight, a condition that was in our favor. We could see 23-millimeter fire near us, but the tracer streams were not close and the SAM crews seemed to be asleep that night. Within a minute or so on our route, I spotted truck lights coming south from Hanoi. At 300 feet above the ground, measured by my radar altimeter, I called my wingman and prepared to release the CBUs. The trucks left their lights on, offering an excellent aimpoint; consequently the bomblets hit at least one truck and started secondary fires, a sure sign of a successful night bomb run. I pulled up to about 1,000 feet and waited for my wingman's attack. About two miles from my location, he saw the secondaries, called his approach, and released his weapons, starting a new fire with his CBUs. That flight was my most successful night interdiction mission out of about fifty night sorties into Vietnam.

My squadron had another exceptional night A-4 mission on 6 March 1968. Two of VA-112's A-4s launched into a black night on an interdiction mission. The overcast was high enough for them to escape SAMs and stay out of most gunfire. The night was so dark that they soon found a string of trucks running with lights on, proceeding south in the Nam Dinh area. Their immediate attack started many secondary fires in the truck convoy. Three or four SAM sites tried to lock onto their airplanes, but only one fired. Dodging the SAM that they saw as its plume filled the night air, the pair also avoided heavy but inaccurate 23-millimeter fire in the area. The two light attack pilots, alone in their A-4s, showed extreme courage while concentrating on their night mission in the face of the threat.

. . .

The gunners at Thanh Hoa Bridge claimed many Navy airplanes during the war, including an F-4 in a 1967 raid by the *Kitty Hawk*'s air wing. The F-4's pilot and RIO ejected safely from the airplane at about 1,000 feet above the ground, but their airplane continued its glide and crashed in the river southeast of the bridge. That F-4 had the latest electronic countermeasures gear installed—ECM equipment that we did not want the North Vietnamese to recover. Looking at the airplane from the air, it was relatively undamaged. The *Kitty Hawk*'s skipper and air wing commander decided to send a few airplanes that night to destroy the F-4 before the North Vietnamese recovered the electronics. Three pilots were selected for the night mission: Lt. (jg) John Lockard, flying a VA-112 A-4, carried parachute flares; two VA-144 pilots carried twelve 500-pound bombs apiece.

It was a clear, starlit night with no moon, increasing the bombers' survivability but making it difficult to spot the downed F-4. The flight catapulted from the *Kitty Hawk* about 2000, joined, and flew the short sixty miles to the beach. Lockard separated from the bombers' section and approached the target area from the south. He wanted to release the flares as planned about 3,000 feet above the river for good illumination of the target. The North Vietnamese had other thoughts. They opened up with radar-guided 57-millimeter fire that Lockard's radar warning receiver detected. Even in the darkness, the antiaircraft bursts were close enough to see, but were drawn off his A-4 by his radar jamming system. As Lockard descended, hoping to mask himself against the terrain, the ground sped by. At his low altitude he could see the terrain and realized that he must be nearing the downed F-4. The river was rapidly approaching his nose, but he was only about 500 feet off the deck, too low for the flares.

Kitty Hawk A-4s approach the bow catapults for a night launch during the Vietnam War. *U.S. Naval Institute*

He pulled up sharply and released six of the flares while going up in what turned into a half Cuban-eight. The flares lit in a curving path following the flight of the A-4, as if the pilot had been in a night air show.

The bombers, meanwhile, had positioned themselves for their bombing runs. Although the antiaircraft gunners were obviously not tracking the three A-4s, flak continued. Luckily there were no SAMs. The bombers had heard Lockard call as he turned toward the river and call again when he released the flares. They were ready to locate the F-4, which soon appeared as an unmistakable shape sitting in shadows near the river. After the bombing run, when there were no secondary explosions from the fuel in the F-4, the flight leader realized they missed the target and ordered a second flare and bombing run.

After Lockard pulled off the target, he had flown out to the river mouth, only a little over one minute's flight time from the bridge. He decided that if a second pass was necessary, he would do the same maneuver but approach on another heading to confuse the enemy gunners. During the second run, flares lit the sky and river again. This time the bomb impacts created a large secondary explosion. The bombers assumed they had hit the F-4 and it had exploded. They declared the mission a success!

Early the next morning, the crew of a photoreconnaissance RA-5 reported that the F-4 was still intact. A few hours later, their photos confirmed this initial report, but the night bombers' damage assessment surely meant that the night strike had destroyed something, maybe ammunition ready to be trucked across the bridge. The expedient but innovative flare release maneuver had not gone for naught. By that time, whatever gear the North Vietnamese wanted would have been removed and the F-4 had become a flak trap, so no further attempts were made to destroy it.[1]

. . .

By November 1965, the North Vietnamese had opened the Ho Chi Minh Trail through Laos into South Vietnam. Almost all military supplies going to Communist units in the south moved along the maze of alternate routes through the Laotian jungle. In periods of inclement weather over North Vietnam, CTF-77 diverted carrier strike sorties, except for the A-6s, to South Vietnam and Laos for troop support or interdiction missions. The large majority of carrier sorties diverted from North Vietnam during the winter went into Laos on "Steel Tiger" missions to interdict the Ho Chi Minh Trail. If the weather turned sour up north, a flight of light attack airplanes—F-4s or F-8s scheduled for a night interdiction mission in North Vietnam—would, just before leaving the ready room to man airplanes, be diverted to Air Force control for missions in Laos.

My war diary includes this entry about a divert to Laos, typical of hundreds of light attack and fighter night missions: "Went on a night Steel Tiger. Flares from an Air Force C-47, 'Lamplighter.' Flares all over the place! Good hits on a truck park. Fair landing aboard on a CCA. Ceiling about 500 feet."

An Alpha strike had been scheduled during that day, but did not go. The weather forecast for the next day was not good, either. Therefore, the air wing commander modified the day's air plan by adding night missions for all squadrons. That night, two sections of VA-112 briefed for separate interdiction missions into North Vietnam, but neither my section nor the other had confidence that the weather would be acceptable for A-4 operations. We were ready to launch as a division of four airplanes when the squawk box told us to divert to Air Force control, probably for a Steel Tiger mission.

After the catapult shots that night, our rendezvous was above an overcast at 20,000 feet on the 220-degree TACAN radial, in the direction of our coast-in point just south of the Demilitarized Zone (DMZ). During the routine join-up and flight to the beach, we kept our navigation lights on bright, allowing for a safer flight into South Vietnam.

Over Dong Hoi, just north of Danang, I reported to the Air Force's air controller. He told me to check in with the target assignment controller in the Airborne Command and Control Center (ABCCC) aboard a C-130 Hercules stationed over Laos. The ABCCC's controller vectored us into the northern part of Laos, where I was told to report to an airborne FAC. The forward air controllers worked twenty-four hours a day, each always assigned to his personal, relatively small sector of Laos, South Vietnam, or Cambodia. It was extremely difficult to spot vehicles at night in the three-tiered tree coverage without assistance from a FAC familiar with a particular part of the Trail. There were no SAMs in Laos or South Vietnam and no 85-millimeter antiaircraft guns. If the FAC and the strike airplanes stayed above 3,500 feet, there was only a small chance of being shot down.

The FAC had found a truck park for our strike group. As we descended into the FAC's general area, we turned off our lights. The flight dropped into trail formation with about a twenty-second spacing between each of us. Ahead, a "Lamplighter" had already started dropping flares, illuminating about a two-hundred-meter square on the ground. Lamplighters were Air Force C-47s that accompanied the FACs at night. They carried hundreds of parachute flares, whereas our light attack airplanes usually carried only six flares along with their bombs. Under the flares, I could see a road running through the jungle in a valley among the karst mountain outcroppings. The FAC fired white phosphorous rockets at the point he wanted us to hit. I could see the dense white smoke that the rocket's warhead made after it hit the ground. The FAC then gave us our dive heading and let us do our job.

I told the flight to set ordnance switches for pair drops. Because there was little threat from gunfire, we each made three runs for increased bomb effectiveness. We released our bombs above the highest flares, which lit about 3,000 feet above the ground. The effect of all the flare light was eerie, but because there was so much light our bombing was more accurate. We relied on radio calls when rolling in and pulling out of our dives to keep each other informed of our positions. I still made it a point to fly the pull-outs entirely on instruments to make the chance of vertigo less likely. After we made our flight's total of twelve runs on the truck park, the FAC pronounced that we did a "Sierra Hotel!" job. According to the FAC, we had placed all our bombs within one hundred meters of the center of the area where he thought some North Vietnamese trucks had parked. We checked armament switches off, turned on our lights, and made a running rendezvous heading back to the *Kitty Hawk*.

As we left, I checked out with the FAC, the ABCCC, and the Dong Hoi controller. By that time we were at 20,000 feet over water. That night we arrived at marshal just in time to make our approach time. When I reported to the approach controller, we received clearance and started the flight's breakup for CCA approaches immediately.

The *Kitty Hawk* had an excellent CCA crew. The CCA controllers guided us to within sight of the Fresnel lens with an accuracy that any land-based GCA unit would envy. The low overcast over the ship was still there, putting me in the soup when I leveled off at 1,200 feet. I lowered the landing gear, tailhook, and flaps, adjusting power and airspeed to maintain level flight for a mile or so. Following CCA's commands, at three miles from the ship I was ready to begin the final approach, pick up the lens, and listen to my LSO.

Even with the lens system, the LSO's guidance, now given over the radio, was valuable in rough seas or low visibility conditions. All carrier airplanes had angle-of-attack (AOA) sensors connected to indicators in the cockpit and in places visible to the LSOs. At night, the LSOs used the lighted AOA indicator to estimate the approaching airplane's speed. Pilots changed attitude measured by the cockpit AOA indicator to adjust approach speed. They changed the power setting to stay on the glide slope indicated by the lens's "ball," reducing power if they saw a ball above the lens's green reference lights or adding power if a low, red ball was visible.

The ship, a dim but distinct blob of light in the blackness, appeared as I descended below the cloud layer. Then I saw the ball. Simultaneously, the details of the deck's centerline lights and drop lights on the stern became visible. Lining up with these lights, which gave excellent lineup references for night landings and making final power adjustments, I believe I trapped on the number two wire. However, if I had had a hook skip or other cause for a bolter at touchdown, the CCA radar controller would have guided and monitored my flight around a low-altitude pattern until I completed a landing.

.   .   .

On another day of bad weather over North Vietnam, I led a division through an area of thunderstorms that stressed my flight's formation-flying skills. After checking in with the ABCCC, we were sent to Cambodia for a FAC's control. The distance to the FAC's position would put us at the end of our radius from the ship for the bomb and fuel load we were carrying. We climbed to about 28,000 feet to conserve fuel, but even at that altitude ran into big thunderstorms. Because of the apparent urgency of the assigned mission, however, I decided to penetrate the storms.

I kept my airplane as steady on course and altitude as possible. In the midst of dense clouds and turbulence, the division was hanging tight to me, but then St. Elmo's fire started. Flying formation while in clouds or at night is one of the hardest tasks an aviator faces. Wingmen easily develop vertigo as the airplanes bounce in the air while they concentrate on the vague shape of their leader, and St. Elmo's fire adds a significant distraction. My wingmen had an increasingly difficult time seeing the adjoining airplanes. The last man in the formation finally suggested we divide. At my command, the division split in the classic instrument condition tactic designed for times when it became impossible to see the next airplane in a formation: The four airplanes spread out with both distance and altitude separation between airplanes. After flying about forty miles, we broke out of the storm with all the airplanes in sight and in the planned positions.

I slowed to allow the division to close and we soon reached the FAC. He was tracking a North Vietnamese troop unit that was about to engage an inferior friendly group. The FAC had us attack immediately. We had good hits on the enemy, disrupting their attack on the friendlies, the FAC reported. By that time, I was already climbing en route back to the ship, barely making the last of the recovery time period. We would have had to tank and stay airborne another hour and forty-five minutes doing nothing if we had missed that recovery window.

. . .

During the first months of 1968, light attack night work over North Vietnam ceased almost entirely while we supported the Marine ground units besieged at Khe Sanh. Around this time, the Air Force and Marines began to use another form of controlled bombing during the night or in overcast weather in northern South Vietnam. The missions, called "Skyspot," aimed at North Vietnamese troop concentrations or supply dumps some distance from the Marines. Controllers using TPS radars positioned flights of airplanes for simultaneous bomb releases by all airplanes in the flight. For carrier aviators, the missions were extremely boring.

Skyspot missions started like any other divert from North Vietnam for the carrier night pilot, but the Dong Hoi controller placed the flight under Skyspot control rather than switching it to an ABCCC and a FAC. A Marine Corps radar site near the DMZ controlled most of the carrier airplanes' Skyspot work. The flight leader followed the altitude, speed, and heading commands of the controller with the precision required for a CCA. The wingmen did nothing but fly a close, parade formation on the leader. There were no SAMs, and antiaircraft fire could not reach 20,000 or 24,000 feet, the altitudes used for the Skyspot bombing runs. The only difference between Skyspot and a peacetime precision approach was that we had to arm our weapons and hit the bomb pickle when the controllers said to release our bombs.

As commanding officer of VA-112, I decided to try something different with the squadron's airplanes—a new tactic to be used without ground radar up North. In the late spring of 1968 as the Khe Sanh campaign ended, I remembered VX-5's A-4 radar modification that had added an adjustable range line on the scope. I studied the bomb ballistics tables to determine the distance from a target at which an airplane flying at 360 knots and 20,000 feet should release a Mk-82 bomb to hit the target. I had the avionics technicians install circuitry for an electronic range line marking that distance in the radar sets of two airplanes. Releasing bombs in a bombing run when the modified radar's range line coincided with the target's radar blip centered on the scope, should result in hits on the target.

With the air wing commander's permission, I picked a North Vietnamese bridge on Route One near Brandon Bay as a test target. On a dark, cloudy night with a wingman, I tried radar bombing with the A-4, leading my section above the overcast offshore, heading toward the target I had selected. The landmark I had chosen as the aimpoint showed clearly on the radar. After centering the aimpoint's blip on the scope, I began tracking it down the center of the display, correcting for wind drift as the aimpoint's image approached the range line, a technique similar to one used by A-6 B/Ns. I had to hold track over the ground, airspeed, and our altitude to hit the bridge. If there had been any SAM alerts on our radar warning system, I would have broken off the attack. We released our bombs as planned when the range line intersected the aimpoint's blip. We had no way of getting a damage assessment of our bombing run, because we could not see the bomb bursts or bridge damage through the overcast. Moreover, weather prevented a photoreconnaissance flight the next day. After discussing the results and risks, the air wing commander and I decided not to pursue that bombing method anymore. It was clear that the A-6 was the only real adverse-weather attack aircraft in the Navy at that time.

What did the A-6A's bombing system have that the A-4 did not have? The A-6s had an INS and a computerized weapon system making a myriad of in-flight calculations to place a bomb on target—plus the A-6s had a second crewman. A-6 crews did practice radar range line bomb releases, similar to what I had done, in case the computer failed, but even with an experienced B/N their hits were consistently poor. The A-4s, in contrast, had no integrated weapon system. Moreover, flying in a hostile environment watching for SAMs and gunfire added immensely to the workload of the A-4 alone in his airplane.

Meanwhile, the A-3 pilot detachments tried to contribute to the carriers' strike effort at the beginning of the Vietnam War by modifying an A-4 bombsight for the A-3. However, that limited them to laydown bomb runs, which exposed the aircrews to heavy gunfire. Very soon, the old, all-weather A-3s performed only refueling or special reconnaissance missions.

▪ ▪ ▪

Task Force 77's night and all-weather attack potential increased in July 1965 with the arrival of the first operational A-6A squadron. VA-75 had finished training and deployed to the war in the *Independence* (CVA-62). The squadron began operations as participants in daylight Alpha strikes, as well as in night strikes. The monsoon started before the squadron finished their deployment and, as expected, the bad weather was perfect for the A-6s. On many days, the A-6s flew the only strikes in North Vietnam by either the Navy or the Air Force.

About 80 percent of VA-75's sorties were at night during its six-month combat tour. At night, the squadron usually flew low-altitude, two-airplane attacks on North Vietnamese installations, losing only one airplane and crew to enemy fire during their five months in combat. That loss was at night to 37-millimeter fire at Bach Long Vi. Unfortunately, during the first month in action, the squadron lost three airplanes and one crew because of ordnance malfunctions. Some bombs' electric fuzes fired prematurely, catching the delivery airplane in its own bomb's blast and fragments. Because the early A-6s suffered heavily from maintenance problems, they flew only a small number of sorties. However, VA-75 showed off the A-6's large bomb-carrying capacity by dropping 25 percent of its air wing's ordnance.[2]

The *Kitty Hawk* brought the second A-6A squadron, VA-85, into combat in late 1965. Although the winter monsoon had started, there were rare clear days, one of which occurred on 22 December 1965, when the air wings in the *Enterprise*, *Kitty Hawk*, and *Ticonderoga* conducted the first strike on a North Vietnamese power plant. About one hundred aircraft in three strike groups, including the A-6s from the *Kitty Hawk*, destroyed the facility at Uong Bi north of Haiphong. However, just three months after that strike, the power plant was operating again. Washington assigned a night strike on Uong Bi to VA-85.

In April 1966, Cdr. Ron Hays led a section of A-6As in the second strike on the power plant. Approaching at low altitude with thirteen 1,000-pound bombs apiece, each A-6's inertial platform tracked the airplane's position in space, not only in latitude and longitude but also in altitude.

Sometimes an A-6A's INS drifted, caused by the gyro's precession, thereby losing its precise position measurements. En route to the target, the bombardier/navigators would correct the system's INS when they sighted recognizable landmarks visually or on radar. In any of the weapons delivery modes, DIANE calculated the airplane's position relative to the designated target with instantaneous measurements of airspeed, altitude, aircraft angle-of-attack, and aircraft range-to-target data. When the B/N selected one of the several different weapon delivery modes, the computer automatically matched the airplane's position and target range data with the correct weapons ballistics tables stored in the computer.

In the night attack on Uong Bi, the crews selected a straight-path, lay-down maneuver in which the system released the bombs at the proper instant in flight to get direct hits on the target. The two airplanes escaped without damage, although there was heavy barrage fire from the defenders' 23- and 37-millimeter guns. Poststrike photography showed that the two Intruders had placed their bomb strings within the boundary of the plant and had leveled the smoke stacks. Radio Hanoi complained about B-52s bombing populated areas in North Vietnam! The A-6 was an impressive and effective airplane when properly used.

VA-85 lost six A-6s during the November 1965 to May 1966 cruise. Their losses were the heaviest of any A-6 squadron during the war. Because a high percentage of the squadron's strikes had gone into the farthest, most mountainous areas of North Vietnam, it is not known whether the losses were from enemy action or collisions with the terrain.

. . .

The *Constellation* and VA-65 relieved the *Kitty Hawk* and VA-85 in the spring of 1966. By that time, the A-6 crews knew that the target track radar, essential to maintaining DIANE's accuracy, was not reliable. If the track radar failed, the system's average miss distance was twice that predicted by the first A-6A's system test results. Despite that disappointment, a single A-6A carrying eighteen 500-pound bombs could achieve a great deal of damage to its target. In addition, the A-6 remained more accurate in combat than the Air Force's F-111s, B-52s, and Canberras, the only other all-weather bombers in the United States's arsenal. Also by that time, the A-6 units had learned that they should check their weapon system and realign the INS after launch by making dummy runs on the carrier before proceeding to North Vietnam. Because fuel was no problem, it was easy to take time for the check.

VA-65, led by Commander Mandeville, experimented with both single-airplane and multiplane attacks. They realized that the INS gave them the freedom to fly singly all the way to a target and still coordinate simultaneous attacks of several airplanes. In contrast, back in the 1940s and 1950s, to perform multiplane coordinated strikes, TBM and AD crews had to fly in formation until near the target before splitting to confuse enemy gunners. VA-65's multiplane strikes were generally composed of two or three A-6s which timed their approaches to release their bombs at the same time-on-target. Usually approaching the target on opposite headings, the crews used the weapon system's straight-path bombing mode to release their ordnance on target. After bomb release, the airplanes turned hard left away from each other, creating separation between them during their egress from the target area. This tactic spread the fire and attention of the defenders and avoided the "shooting gallery" situation created by a bomber stream on the same heading. There were no VA-65 airplanes hit during any of their multiplane strikes.

During Rolling Thunder, A-6 crews planned their missions to destroy specific, assigned targets. At this time in the war they did not consider any of their night flights as interdiction missions offering chances to hit and destroy targets of opportunity. VA-65's Lt. Cdr. Pete Garber and Lt. Don Hahn destroyed a railroad ferry slip bypassing the Thanh Hoa bridge on one significant, preplanned, single A-6 night mission. After confirming that a large number of railroad cars were stranded on the north side of the river, the air wings from two CVAs struck and destroyed many boxcars, tank cars, and tracks the next day, an incident reminiscent of Korean War actions.[3]

The *Kitty Hawk* and *Enterprise* with VA-85 and VA-35, respectively, were back operating during the 1966/67 winter monsoon. VA-35 was the first deployed A-6 squadron based on the West Coast. The A-6A's maintenance record improved, allowing effective use of the night and inclement weather tactics tested during the first two deployments of the airplane. However, the A-6, because of its large payload, also continued to be in demand for daylight Alpha strikes. Having to hold airplanes for the Alpha strikes greatly reduced the squadrons' night work.

The *Constellation* again relieved the *Kitty Hawk* in late spring of 1967. Throughout the summer, her A-6 squadron, VA-196, flew with the rest of the air wing, executing large daylight Alpha strikes. When free of the daylight missions, though, the squadron tried multiplane coordinated night strikes. The executive officer of VA-196, Cdr. Robert Blackwood, led a multiplane night strike into the Hanoi area on the night of 27 October 1967. He and his B/N with five other Intruders proceeded at low altitude west through the Red River Valley to strike the Hanoi ferry slips, a large supply area.

The A-6s planned for at least a thirty-second time separation between aircraft at the target to prevent midairs and escape bomb blasts. Going in first, Blackwood released his bombs and escaped through the center of the Hanoi defenses, meeting little antiaircraft fire; the enemy fired no SAMs at him. However, as the A-6 flight streamed through the defenses on the same heading taken to the ferry slip by Blackwood, not surprisingly the gunfire and SAM activity increased. The last of the six A-6s had at least six SAMs fired at it, but there were no losses. In contrast, Bill Martin's 1945 multiplane TBM strikes consistently lost more than one airplane per strike. The Japanese defenses must have been tougher or the slower speed of TBMs did not allow them time to escape. Nevertheless, because of the increase in antiaircraft gunfire at the last A-6s in Blackwood's strike group, the Task Force 77 staff and the VA-196 skipper decided that multiplane strikes were not safe, ignoring the earlier success achieved by VA-65.

Three nights later, Lt. Cdr. Charlie Hunter and his B/N, Lt. Lyle Bull, launched from the *Constellation* with eighteen 500-pound bombs to strike the same Hanoi ferry slip. The defenses were ready after the previous multiplane Intruder strike. About twenty miles from the target, the SAMs started—Hunter and Bull saw and avoided about six. Even at 100 feet above the ground, while Hunter evaded SAMs with vertical weaves— at one point completing a barrel roll—Bull managed, with the aid of DIANE, to place the aiming cursor on the target. In the immediate target area, searchlights came on, sweeping the night sky trying to pinpoint the Intruder for the 23-millimeter batteries. The fire-hose tracers were wildly inaccurate but were visible and distracting to Hunter and Bull while they concentrated on the "pathway in the sky," radar scope, and countermeasures equipment. Popping up to 1,200 feet for the final bomb run, Hunter held the A-6 straight and level for the required several seconds. After DIANE released the bombs on the ferry slip, there was no time to look for bomb damage. Hunter turned sharply and started jinking up, down, and sideways, leaving Hanoi behind him. The mission was over when he and Bull landed aboard the *Constellation* after a full instrument procedure approach.

. . .

168

Six of the ten A-6 deployments during the Rolling Thunder campaign covered the winter monsoon season, making full use of the A-6's all-weather systems. The *Kitty Hawk* returned for the third time with a squadron of A-6s during the 1967/68 winter monsoon. The *Kitty Hawk's* commanding officer, Capt. D. C. Davis, asked for an all-Intruder air wing for the 1967/68 deployment, but that was not possible—there were not enough A-6s! If there had been more A-6s in the Navy inventory, the large carriers operating during the winter monsoon could have been more effective with a full load of those all-weather airplanes rather than only one squadron.

Cdr. Jerry Zacharias led VA-75 in the *Kitty Hawk* on that deployment. The squadron had been to the Mediterranean in the *Independence* the previous year; therefore, few veterans of the squadron's first Vietnamese tour remained. His squadron's record was typical of an A-6 unit's Rolling Thunder effort during the adverse weather season, flying about 1,500 sorties, 20 percent on day Alpha strikes, 41 percent in daylight inclement weather conditions, and 39 percent at night, in clear or instrument conditions.

Although VA-75 did fly a few two- or three-airplane night strikes, most of their missions were single-airplane strikes. On one night multi-plane operation, five airplanes simultaneously hit four different targets. The mission coincided with the first use of the large Standard ARM anti-SAM missile for SAM suppression during the attack. The two airplanes striking the same target used widely separated approach routes and attack headings to avoid the danger caused by using the same bombing run heading, but arrived at the target within seconds of each other.

As the *Kitty Hawk* left the Gulf of Tonkin for a rest period in February 1968, the NCA scheduled an important strike on the Hanoi port facilities. VA-35 in the *Enterprise* had already relieved VA-75, but because of the importance of the strike and because this was VA-35's first line period of a long deployment, Commander Zacharias took two of his airplanes to the *Enterprise* to lead the strike. VA-35 added two A-6s to the strike. Zacharias planned that each of the four airplanes would release their eighteen Snakeye bombs in laydown runs across the target at two-minute intervals.

Scheduled for takeoff at 0100, 24 February, only three of the A-6s made it off the deck; one had radio failure before launch. After they were airborne, Zacharias and his B/N, Lt. Cdr. Mike Hall, determined that their INS had not been properly aligned. Zacharias climbed to the A-3 tanker's position using the small backup attitude indicator. The refueling did not go well. The tanker had an electrical failure that caused the refueling pumps to stop midway through. Without the extra fuel, Zacharias and Hall would have barely enough fuel for the flight to Hanoi and back to the *Enterprise*.

After leaving the tanker, Zacharias and Hall took thirty minutes to realign their INS, providing them with a functioning navigation and weapon system for the strike. However, because of this delay, they were late arriving in the Red River Valley. The VA-35 A-6 had reached the target thirty-five minutes earlier, released its bombs, and departed the area. The other Intruder had released its bombs miles short of the target, perhaps because of a similar INS malfunction. Now, Zacharias and Hall were the second of two widely spaced single-airplane attacks on the port facility rather than part of the planned four-airplane coordinated strike.

Cdr. Jerry Zacharias briefs his squadron for the night's strikes. *U.S. Navy, courtesy Capt. J. M. Zacharias*

170

Zacharias had crossed the coastline south of Ninh Binh and flown west and north, circling around Hanoi over the shadowy karst mountains, waiting until he was west of the city to turn toward their target. The night was silent in the mountains, with just the drone of the engines in their ears. The pitch black night outside the cockpit—no lights on the ground, no moon—contrasted with the glow of the VDI and reflections from the radar scope and other lights inside their A-6. At the final checkpoint over the mountains, Zacharias turned east toward the flat Red River delta and commenced a descent to 200 feet while accelerating to 450 knots. During descent, their radar warning sensors set off audio signals indicating that the first SAM radars had begun to paint their airplane. Closer to the target they heard indications that the enemy had many antiaircraft gun radars aiming at them. There was almost too much radar warning noise in their earphones to make sense of what might be coming at the airplane.

Leveling off at 200 feet still on an easterly heading, they saw the bright flashes of two SAMs lifting off pads directly in their path. Immediately afterward, the two night aviators were blinded by the brilliant red missile warning light in the cockpit. Looking out of the cockpit, Commander Zacharias saw the missiles' rocket plumes reflecting off the flooded rice paddies. The reflections were bright enough to act as an attitude reference as he made a turn away from the missiles and dropped to 100 feet. He forgot about the speed limitations on the Snakeyes and put the throttles full forward. Zacharias had gotten so low, farmhouses whizzed by the cockpit in the ghostly light. Hall warned that they were at 50 feet! As he jerked the A-6 back to 100 feet, Zacharias dropped chaff, banked 85 degrees to the right, and pulled five g's to avoid the SAMs. One missile exploded and shook the airplane violently. Although the crew thought that both missiles had missed, their plane captain back aboard the *Enterprise* would later find a shrapnel hole in the wing just forward of a main fuel tank.

With the missile threat eased for a moment, Hall gave Zacharias the final run-in heading to the target. He, as B/N, had been tracking the port facility on radar all through the missile evasion maneuvers. As they neared Hanoi, there was lots of antiaircraft fire. Zacharias later said it looked like the pictures of Baghdad during Desert Storm, only more so. He and Hall were the last on target, and it seemed as if everyone in

Hanoi was awake and shooting. Antiaircraft fire was ahead and to the right and left of their A-6, but not in the target area. From the antiaircraft artillery's illumination, Zacharias could see the bends in the Red River near the port facility. He increased altitude to 500 feet and they dropped their eighteen Snakeyes on target. Zacharias broke left across the center of Hanoi, maintaining 400 feet while flying in between the flak sites. The Vietnamese antiaircraft gunfire was all barrage fire, not aimed—the gunners shot straight up, hoping the A-6 would fly through the massive screen of bursts.

As Zacharias and Hall were en route to the ship that night, the North Vietnamese fired another missile at the A-6, but because of their missile evasion tactics, they arrived at the ship with no further harm. They landed with only 1,100 pounds of fuel remaining (normally the A-6s returned with at least 2,000 pounds in their fuel tanks).

Most of the gunfire hits on the A-6s came from barrage fire. Zacharias knew that multiplane strikes were dangerous because once the first strike airplane alerted the antiaircraft gunners, they started a continuous, unaimed barrage fire over the target when they heard another airplane approaching. Zacharias clearly preferred the night for bombing. He felt that the chances of survival were much greater in a night bombing strike than in a day Alpha group.[4] Statistics indicate there was some increase in survivability at night. There were twenty-five A-6s lost by the end of 1967: fourteen on day strikes, eight on night strikes, and three in operational accidents.

Commander Mandeville added that nothing was more disconcerting to an A-6 crew than to be hiding in the middle of a cloud layer on a single-airplane day strike and have the clouds open. He was caught in that situation once, but he and his B/N managed to escape the ground fire after the Vietnamese gunners spotted their airplane. Early in the war, the A-6s relied heavily on their electronic countermeasures system and on low-altitude approaches to avoid enemy fire.[5]

On 31 March 1968, U.S. air activity over most of North Vietnam terminated for almost four years when President Johnson declared a bombing halt north of the 19th parallel. Only the coastal panhandle area between the DMZ and latitude 19 degrees north remained an open target area in North Vietnam. The president gave the on-scene commanders authority to strike all military facilities and the transportation network in that region at their discretion.

CTF-77 began a concentrated air interdiction campaign, the "Panhandle Campaign," to stop all road and trail traffic in that narrow sector of enemy territory that was his responsibility. The task force's daily air plan scheduled continuous cyclic operations from the two or three carriers in the Gulf of Tonkin. If there were two carriers in the Gulf, one carrier conducted flight operations from 0600 to 1800 and the second carrier flew all night. If there was a third CVA on line, then the second CVA flew from noon to midnight and the third from midnight to noon while the first continued the 0600–1800 schedule. Because all the aviators were night-qualified, all carriers now took their turn as dedicated night carriers. That was a situation much different than in World War II, when only four—never more than two at a time—night CVs operated with night air groups.

The air operations officers divided the flying day into an appropriate mission cycle for their type of airplanes, with at least four cycles at night. There were at least four light attack sections, four fighter sections, and one section of A-6s, if in the air wing, flying in each night mission cycle, saturating their part of the free-bombing area in North Vietnam. At the end of the summer, CTF-77 reported that traffic through his area was reduced by at least 85 percent since the beginning of the Panhandle Campaign.

VA-94 was particularly successful during that period. Practicing section tactics with or without flares, if one of the squadron's A-4s detected trucks or truck lights, the pilot attacked and expected secondary explosions or fires from the strafing or bomb run. Then other A-4s or A-6s in the area would concentrate on the fires to destroy any remaining trucks.[6]

That summer, about one-third of VA-23's flights were on night interdiction missions, which were usually successful in destroying enemy trucks. The squadron achieved success at night using section, low-altitude attacks with Mk-82 Snakeye bombs. Their air wing strike planners in the *Ticonderoga* studied bomb damage photos taken during the day, selecting a downed bridge along one of the truck routes into South Vietnam. The night flyers then planned to bomb trucks stopped and backed up on the north side of the damaged bridge. Typically on a night mission, a section of A-4Fs flew in close formation, remaining at low altitude after rendezvous. Timing a short flight inland from the coast to Route One, the flight would turn ninety degrees to the north. The turn put them looking into "cat's eyes," the half-headlights of enemy trucks along Route One. At an altitude of 1,000 to 2,000 feet above the flat terrain, the attackers could usually find trucks stopped at the downed bridge, just as they planned. One quick pass dropping Snakeyes would usually set secondary fires and light up the target area for succeeding flights.[7]

VA-75 in the *Kitty Hawk* was still on Yankee Station when the Panhandle Campaign started. The A-6A had a good Doppler automatic moving target indicator (AMTI) system. With that sensor, an A-6 could spot moving trucks and hit them with bombs or CBUs. On some missions a section of A-4s would be in the vicinity and would hurry to the scene of the A-6s' attack, if secondary fires started. VA-75, using their AMTIs in cooperation with the A-4 squadrons during the early part of the Panhandle Campaign, destroyed several trucks.[8]

The A-6s of VA-85 in the *America* (CVA-66) also decided to try the team concept. They used their AMTI systems and teamed with VF-33's F-4s loaded with Snakeyes to find and destroy moving trucks along the panhandle routes. The *America's* air wing did not try that tactic until near the end of the campaign, however, so it was not clear whether the A-6 and F-4 teams achieved a greater success rate than would have been achieved with A-6s and F-4s flying and destroying the enemy separately.

During the middle of 1968, few SAMs operated in the panhandle area and gunfire was sporadic and ineffective. Although the attack and fighter airplanes flew many sorties, losses during that period were few.

The first trial of side-looking radar—air-to-ground acquisition systems for night work—occurred during the Panhandle Campaign. Theoretically, these systems had better radar resolution and would give a clearer picture of small objects such as trucks. There were two different side-looking radar systems: one on the reconnaissance version of the A-3, the RA-3, and another on the Army's OV-1 Mohawk. The RA-3s were normally based at NAS Cubi Point, but could fly from a carrier. The Mohawks flew from an Army field in South Vietnam. Both units coordinated their night work with the carrier airplanes through CTF-77's staff. Teams of an RA-3 or Mohawk and two light attack airplanes would arrange to join at a TACAN point off the North Vietnamese coast. The search airplane would then search along a stretch of Route One, attempting to acquire truck traffic. If the side-looking radar detected trucks, the RA-3 or Mohawk controller would vector the light attack airplanes to the trucks using TACAN positions or preplanned coded geographical positions. Unfortunately, the night missions flown by RA-3s or OV-1s teamed with carrier airplanes were not as successful as those of the A-6s using AMTI.

. . .

On 1 November 1968, President Johnson ceased all bombing in North Vietnam, ending the Panhandle Campaign. During the next three and a half years, no carrier aircraft flew over North Vietnam, although nine carriers deployed on eighteen tours with Task Force 77 in the Gulf of Tonkin. Those carriers and their air wings launched about 120 strike sorties each day supporting troops in South Vietnam and the interdiction campaign in Laos. The A-4s, and later the A-7s, made up the bulk of those sorties.

Night work in Laos was always interesting because of the number of airplanes, not just the Navy's, flying without lights through the karst mountain valleys. FACs continued to control the airspace, but there were so many bombers that at times bombing patterns overlapped. The FACs found a way other than flares and white phosphorous rockets to mark targets at night: They dropped something known as a "log" that burned for many minutes on the ground. The FACs selected aimpoints relative to the position of the log for their strikes.

Not long after the bombing halt, in cooperation with the Air Force, CTF-77 decided to use the A-6's radar bombing systems in Laos. The Air Force had been radar bombing in Laos with its F-111s even before the bombing halt in an operation they called Commando Nail. To support that mission, their crews had been gathering radar scope photography of road intersections, bridges, and other locations that might be good interdiction points along the Ho Chi Minh Trail. I, at that time a member of CTF-77's staff, took Lt. Lyle Bull with me to the Air Force's targeting headquarters, where we arranged for the transfer of radar target photos of possible targets for which our A-6s would be responsible. Compared to our work during the 1950s in VC-35, the quality of radar scope photography had advanced more than 100 percent. After a few weeks, the

A-6 squadrons had a good library of photos and were ready for all-weather missions along the Trail. The missions that the A-6s flew on their own into Laos during the temporary bombing halt kept them in practice for the riskier work still ahead of them in North Vietnam.

A modification of the A-6 appeared with VA-165 in *America* in April 1970. The A-6C TRIM (Trails and Roads Interdiction Multisensor) entered the war with the promise of using its forward-looking infrared (FLIR) and low-light-level television systems to find those small, elusive truck targets along the Ho Chi Minh Trail. VA-145 in *Ranger* also deployed with the A-6C later that year. The squadrons participated in night "Commando Bolt" operations controlled by the Air Force from their command center at Nakom Phenom, Thailand. When the Air Force ground sensors detected truck movement along the Ho Chi Minh Trail, flights of aircraft would be allowed into the target area for search-and-attack bombing runs on the trucks.

Although state of the art, the A-6C's infrared systems were not yet fully developed; therefore, the systems were not that effective. Furthermore, often the trucks stopped and hid under trees, eliminating the chance of FLIR or radar detection. It was then that the old, reliable FACs offered the opportunity for dive bombing attacks on trucks they could find visually.

The A-7A Corsair II, introduced in 1967 as the A-4's replacement, was soon itself replaced by the A-7B with a more powerful engine. The A-7B had about one and a half times the payload and endurance of the A-4, making it a much-preferred carrier airplane. Like the later versions of the A-4, it had an air-to-ground radar and a computer-aided weapon system, but the A-7B was still a visual attack airplane, not specialized for night attack.

On the other hand, the final Navy version of the A-7, the A-7E, was a revolutionary light attack airplane with a fully computerized bombing system that made the A-7E more accurate than earlier light attack airplanes. It was the first light attack airplane with an INS. It was also the first Navy airplane with all flight information displayed not only on the instrument panel but also in a head-up display (HUD). The A-7E had a radar that could distinguish and mark radar-significant objects as small as concrete and steel bridge pilings. The bombing system, together with the INS for navigation and radar for target acquisition, made the A-7E a

A *Ranger* A-6 ready for its night mission into North Vietnam.
U.S. Naval Institute

good night attack airplane. The pilot could select a radar aimpoint, place the weapon system cursor on the radar blip, and let the weapon system automatically release bombs at the proper release point for accurate hits. In that mode of operation, with the A-7E flying at 10,000 to 20,000 feet, the airplane was almost as good a radar bomber in combat as the A-6A. The first squadron of A-7Es appeared aboard the *America* in Southeast Asia during April 1970.

.   .   .

During 1969, the North Koreans shot down a U.S. EC-121 on patrol over the Sea of Japan. The *Bon Homme Richard* went with her air wing, CVW-5, to the Sea of Japan ready to conduct a punishing, one-time strike on a selected North Korean airfield in retaliation for the aggression. Cdr. A. A. Schaufelberger, the air wing commander, and his squadron commanders planned the attack. Because there was an overwhelming number of North Korean day fighters—about four hundred tallied by our intelligence—but no North Korean night fighters, Schaufelberger decided to strike at night with his light attack airplanes, escorted by F-8Js in the event that the MiGs did attempt to disrupt the night strike.

The plan called for Schaufelberger to lead a division of A-4s and Cdr. William H. Robinson, commanding VA-144, to lead another division. The two leaders had had more radar experience than any other pilots, Schaufelberger in VC-33 and Robinson in VC-3. They planned to make landfall aided by the A-4s' radar at different points on the coast. Then, approaching the airfield, recognizable on the A-4s' radar scope, the strike group would totally surprise the defending gunners in the middle of the night and heavily damage the airfield.

The final execution order to conduct that night strike never came.[9]

. . .

The North Vietnamese tried to end the war in April 1972 with another invasion stronger than the 1968 Tet Offensive. Twelve of thirteen North Vietnamese divisions were engaged in South Vietnam by 6 April. As Marine Corps units returned to South Vietnam to aid the South Vietnamese army, the first series of Navy air strikes provided direct and close air support to the ground units in the front of the battle. Three CVAs in the Gulf launched 4,800 sorties day and night against North Vietnamese troops in the South during April. The Air Force FACs continued to control the carrier airplanes, and Skyspot directed about 80 percent of the carriers' night troop support missions.

Within a week after the invasion, President Nixon rescinded the bombing halt on the North. The NCA authorized strikes on military facilities in the panhandle and on special targets farther north. On 16 April, the president unleashed B-52s on prime North Vietnamese targets. Nightly bomber streams entered North Vietnam through the Laotian border flying at 35,000 feet, struck Hanoi, Haiphong, and other major cities, and exited over the coast into the Gulf of Tonkin. On the first night, the *Coral Sea*, *Constellation*, and *Kitty Hawk* launched fifty-seven light and medium attack aircraft armed with bombs and Shrike and

Standard ARM missiles to provide SAM suppression for the B-52s. Every night that the B-52s were in the air, these aircraft covered every known SAM site across the Red River Valley. SAM weapon system activity was so heavy when the B-52s appeared that the SAM radar sites made easy targets for the Shrike and Standard ARM shooters.

By May it was certain that the Army and Marine Corps units had stopped the North Vietnamese army in the South. Consequently, the carriers gradually shifted their sorties from troop support missions to interdiction missions and strikes on fixed targets in the Hanoi and Haiphong area. President Nixon finally authorized the mining of Haiphong Harbor on 8 May. A-6s from VA-165, VA-52, and VA-115 in the *Constellation*, *Kitty Hawk*, and *Midway*, respectively, mined the major harbors at Haiphong and Cam Pha, as well as laying smaller mines in some of the rivers. The Intruders performed the initial mining operation because their inertial navigation and radar systems could accurately mark mine locations.

B-52 night strikes intensified on 10 May, beginning the Linebacker I campaign that lasted into October. The Linebacker I strikes hit most facilities of possible military or economic value in North Vietnam. Task Force 77's primary mission was to stop North Vietnamese traffic to the South. The task force started flying cyclic schedules again in May just as it had during the Panhandle Campaign, with three carriers keeping airplanes over North Vietnam twenty-four hours a day. The carrier air wings flew about 30 percent of their 12,000 strike flights the summer of 1972 on interdiction missions at night. The light attack squadrons split their time evenly on day and night missions, while the eight A-6 squadrons in the large carriers spent about 40 percent of their flight time on night interdiction missions. VA-35 in the *America* had A-6Cs, but again the TRIM aircraft were only marginally effective in locating North Vietnamese trucks.[10]

In periods of inclement weather as the winter monsoon moved in, the A-7Es joined the A-6s over the North as radar bombers. The light attack squadrons not equipped with A-7Es again diverted to South Vietnam and Laos. The A-7Es used their all-weather systems on bridges and radar-significant chokepoints along the North Vietnamese road network in the panhandle. Almost as precise as the A-6 system, the single-seat A-7Es had better results than those I achieved experimenting with radar bombing in the A-4C a few years earlier.

During preflight before a radar bombing mission, A-7E pilots entered the target coordinates into their weapon system computers. En route to the target, the computer predicted the heading to the target with inputs from the INS. When the target blip showed on the radar's map display of the ground, the pilots would update the inertial position. Retaining section integrity, the A-7E flight leader centered the target blip on radar and set up his weapon system while giving instructions for his wingman to follow suit. The leader then simply followed the weapon system's indicator needles until the system released the bombs. The leader gave the simultaneous release command to his wingman as his system released. SAM firings were almost nil, probably because the North Vietnamese were saving what few they had for future B-52 strikes; therefore, it was relatively safe for the A-7Es to fly above the cloud tops at 20,000 feet. Moreover, at that altitude, only a few 85-millimeter guns could reach the A-7Es, and their bursts were inaccurate.[11]

The North Vietnamese air defense system concentrated on stopping the Navy's air attacks with antiaircraft guns and SAMs, although there was air-to-air action between the large day strikes' escorts and MiGs. The MiGs did not fly at night except for a few attempts to intercept the B-52s, but it was during Linebacker I that a carrier fighter finally destroyed an enemy at night over land.

Shortly after twilight on 20 August 1972, Lt. Cdr. Gene Tucker and his RIO, Lt. (jg) Bruce Edens, of VF-103 sat in Condition Five alert in the *Saratoga* (CVA-60). They were in one of the latest fighters, an F-4J with an improved weapon system. The AWG-10 had the APG-59 Doppler radar, the Navy's first look-down radar that enabled detection of airplanes at low altitude. CTF-77 ordered Tucker and a VF-31 aircraft to launch in response to a bandit detected over land near Vinh.

After a few minutes following erratic vectors, Edens detected the bandit flying on a northerly heading about twelve miles inland. Tucker hit burner and accelerated to over 600 knots. His RIO lost radar contact at about four miles while they were overtaking the bandit, as the MiG descended to an altitude just above the 4,000-foot-high karst to the west of Vinh. Tucker turned port, slowed, and on a hunch descended, being careful to go no lower than 4,000 feet; he then turned north again parallel to the bandit's course. Edens regained radar contact and again Tucker accelerated. He drew within three miles astern of the bandit before firing a Sparrow. Blinded by the Sparrow's rocket motor when it launched, Tucker waited five seconds to fire a second Sparrow. Then a tremendous explosion lit up the night. Tucker broke hard starboard to avoid wreckage he knew would be directly ahead of their F-4. In the war's nine years, that was the third and last airplane destroyed at night by Navy F-4s, the only Navy fighter to get into night air-to-air action.[12]

. . .

Linebacker I ended in October 1972 when the North Vietnamese appeared ready to retreat from South Vietnam, but the carriers continued cyclic strike operations into the panhandle region, plus infrequent strikes into the Nam Dinh and Ninh Binh "hourglass" area throughout the fall. When the North Vietnamese showed they were not ready for peace by continuing to fight in South Vietnam, Nixon ordered Linebacker II.

The operation started with large B-52 raids on Hanoi and Haiphong on 18 December 1972 and lasted through Christmas. The Navy's light and medium attack squadrons flew nightly ECM and SAM suppressor sorties during the B-52 stream attacks. The Shrike and Standard ARM–equipped aircraft were supposed to hit the North Vietnamese SAM sites thirty minutes before the arrival of the bomber stream, but they did not have accurate location information about the mobile SAM sites. The suppressors' success was due to their withholding fire until the sites radiated, then shooting their antiradiation missiles.

Foul weather kept many scheduled Navy Alpha strikes from launching, but despite the weather, the A-6s continued nightly strikes with single-airplane attacks on key targets.[13] The B-52 night operations climaxed on 26 December 1972 with a one-hundred-airplane raid in three streams protected by Navy jammers, SAM suppressors, and night fighters. After the B-52 strikes ended on 29 December, the carriers began attacking the panhandle road network again, but these day and night strikes continued only for another month. The *Oriskany* flew the last carrier strikes of the Vietnam War on 27 January 1973.

After the North Vietnamese signed a cease-fire agreement in 1973, the CVAs and their air wings reverted to their prewar missions, although with less emphasis on nuclear strike operations than they had had between the Korean and Vietnam Wars. Although the reductions made in the Navy's structure at the close of the Vietnam War were not as drastic as those in 1945, a new austerity program did bring a smaller force. When the antisubmarine carriers decommissioned in the early 1970s, the sea-based fixed and rotary wing ASW squadrons found new homes in the redesignated attack carriers, again called CVs, or CVNs if nuclear-powered. As the number of carriers came down, the Navy reached another milestone in 1975 when the *Nimitz* (CVN-68), the second nuclear-powered carrier and the first of a large class of nuclear-powered carriers, made her maiden cruise. Furthermore, the Navy equipped the carriers with the latest, state-of-the-art airplanes, all maintaining a night and all-weather capability.

NATC's night and all-weather syllabi for prospective carrier pilots and NFOs showed little change in the twenty years between the end of the Vietnam War and the end of the Cold War. One exception was that while in flight training, the student jet pilot again received the Navy's standard instrument card, a practice renewed from the 1950s.[1] Further, in the 1980s, as the Navy introduced other sensors such as infrared and complex ECM systems into the fleet, the student NFOs received more classroom work in electronics theory.

When the student pilots and NFOs received their wings throughout the 1970s until the summer of 1985, the Navy assigned them to fleet squadrons, but they continued to attend a course at an instrument FRS. During the summer of 1985, because the instrument training syllabus in NATC had developed into a complete course, the Navy disestablished the instrument FRSs. Thereafter, reporting directly to a fighter, medium attack, or light attack FRS, the nugget pilots, RIOs, and B/Ns started training in the appropriate airplane's weapon system simulator.

After completing ground school, the students began a short instrument flight course in their type of airplane. All FRSs scheduled their basic instrument navigation flights at night as they had been doing during the Vietnam War, but as the flight simulators improved through the 1980s, the FRSs' reliance on flight simulators for teaching flight techniques increased. The flight simulators up to 1980 were better than the Link trainer but were still primitive, providing practice only in emergency and basic procedures for the type of airplane's flight and weapons systems. Because the first simulators' electronic systems often malfunctioned, instructors were left with nothing but lecture material for their students. By 1990, however, flight simulators had improved dramatically, to the point that students flew all basic instrument flights in a simulator, reducing the syllabus night time requirements. To build up their actual night flying experience, the students flew about 60 percent of the aerial refueling training sorties at night.

The fighter squadrons emphasized air intercept practice in their night tactical work, flying about one-quarter of their intercept training flights at night. The procedures for night and day intercepts were the same, but there was one significant difference between day and night tactics. At night, only single airplanes, or rarely two airplanes in a "welded wing" section, conducted intercepts. The fighter RIOs flew just a few flights with an instructor pilot, then joined the nugget pilots for the remainder of their training syllabus. The night F-8 gun fighters learned the basics of night intercepts with a stern conversion (fish hook) maneuver to get into gun firing position. These new pilots experienced much of the excitement and uncertainty of closing to gun range that earlier night gun fighter pilots had felt.

The fighter FRSs also flew a limited number of day and night navigation and air-to-ground weapon delivery sorties, expecting to perform night interdiction missions. The light attack FRSs flew 25 percent of their long tactical navigation sorties at night. During the weapons training portion of the syllabus, the A-6 crews and A-7 pilots learned to use all modes of their respective weapon delivery system at night, as well as visual dive bombing techniques over flares or without flares. A-6 crews were aided by instructor pilots or B/Ns for the first few flights, then, like the F-4 and F-14 crews, began teaming with a new pilot or B/N. A-6 pilots and B/Ns flew about 45 percent of their syllabus at night.

. . .

Pilot and flight officer training ended in the fleet replacement squadrons with periods in which they learned the techniques of carrier landings, first in the simulator, if one was available, and then during field practice. The flight syllabi scheduled only a few day FCLP (now called mirror landing practice, or MLP) sessions, some of which were added to the end of other flights. The new flight officers flew a few carrier landings with instructor pilots to learn their new duties during an approach and landing, but did not fly with new pilots during their first carrier landing attempts in the operational airplanes.

In 1972, the Navy introduced the A-7 Night Carrier Landing Trainer (NCLT) at NAS Cecil Field. The simulator re-created the darkness, the ship's lighting, and the sensations of flight that the A-7E pilot saw and felt when approaching the carrier for landing. The first A-6 NCLT appeared in 1982. After the introduction of carrier landing simulators, coordinated with actual MLP, carrier night landing safety improved dramatically. As an example, after the introduction of the A-6 carrier landing trainer, the A-6 accident rate was cut in half.[2] Full flight and weapon system simulators for each type of operational airplane now allowed pilots to practice carrier landings, tanking procedures, air-intercept, and air-to-ground weapon tasks. These realistic simulators made possible a reduction of flight hours for training the students, but still allowed production of skillful night and all-weather aircrews.

By the early 1970s, all approaches to a carrier at night or in adverse weather conditions were in accordance with instrument flight procedures. Because all pilots were proficient in precision approaches to a landing, formation approaches were eliminated. The CCA system evolved during the 1970s and 1980s into an acceptable automatic landing system. The first operational element of the new ACLS was the APC introduced in the Crusader. An APC interpreted speed commands transmitted through an ultra-high frequency data link from the ship's precision radar. The F-4 was the second airplane to use an APC for landings, followed by the A-7E. The A-6 was the last carrier airplane of the 1960s generation to receive an APC. The second part of the CCA system was an instrument landing system (ILS) used to track approaching airplanes. The ILS fed glide slope and heading information through the data link to indicator needles on the pilot's panel to guide him down the final approach path without voice radio transmissions.

By 1990 there were four modes offered in a carrier-controlled approach. The modes ranged from full automatic control of the airplane to controller talk-down procedures. Weather condition minimums varied with the CCA mode. After becoming familiar with the ACLS, most pilots used Mode II, in which the system controlled speed but the pilot followed system glide slope and heading information provided on his instruments. Mode II minimums allowed the pilot to fly to 200 feet or in sight of the lens at three-quarters of a mile from the ship. Pilots relied on that mode of the ACLS at night and in bad weather to augment their skill, but distrusted full ACLS control because the system was unreliable.

There was one exception to the rule banning visual or formation approaches at night or in adverse weather. If an airplane had a radio or an electrical failure in flight, the disabled airplane's wingman led it to within sight of the lens. The pilot with the radio or electrical failure then flew the landing approach to arrestment. If the "no radio" aircraft did not catch a wire, then the CCA controller kept the landing pattern clear until it landed successfully or diverted to an airfield ashore.

One of many interesting no-radio situations on dark and stormy nights started as a double-cycle (3.5-hour) flight for a division of A-4s with the pilots wearing cold weather survival suits for flights over the frigid North Atlantic. The weather was marginal at launch: cloud tops at 31,000 feet, bottoms at or below minimums. Half an hour into the flight, all three wingmen had lost their radios. After rendezvousing the division on top, the leader finally started down through the dark clouds. One pilot got vertigo, but all hung together. Fuel was not a problem and the leader released each of the no-radio airplanes on the ball, after which each pilot made a perfect landing. The excellent procedures of the air wing, CVW-8, for emergencies and system failures had been thoroughly briefed before the flight and were carefully adhered to by each flight member. Since there was no divert field because of changing weather at the scheduled airport, those standard procedures probably saved a life or two.[3]

There was no question that by the end of the Vietnam War, the fleet had completely absorbed the pilots who had gained night experience in the composite night fighter and attack squadrons. They had transferred their knowledge to the younger carrier pilots, all of whom were night and all-weather qualified. Although the night pilots of the earlier era did not need to pass on the challenge of bringing an airplane back aboard an ill-lighted, straight-deck, barrier-equipped carrier, a carrier night landing was still a stressful experience for carrier combat pilots. Flight psychologists from the Navy's Aeromedical School had conducted an experiment during the Vietnam War in which flight surgeons placed heart sensors on a number of carrier aviators flying in combat. The results of that test showed that all phases of carrier night operations, including the need to be on time at marshal, the catapult shot, and the landing, produced more indications of stress than the heaviest combat.

. . .

The Navy introduced the F-14 Tomcat in January 1973 at the end of the Vietnam War, and it remains active. The Tomcat is a large, two-crew, twin-engine, night and all-weather fighter. The F-14 can achieve supersonic speeds, but at those speeds, its high fuel flow makes its range and endurance extremely short. The F-14's primary air-to-air weapon is the AIM-54 Phoenix, a long-range, radar-guided missile. The F-14 also carries the infrared-guided Sidewinder and a Gatling-type 20-millimeter cannon. Optionally, it can carry the Sparrow or the later AIM-120 AMRAAM in place of the Phoenix. Early versions do not have an air-to-surface weapon system.

The F-14 instrumentation includes two cathode ray displays similar to the A-6's instrumentation. It has a vertical display indicator that presents attitude and steering information to the pilot. The picture on the VDI resembles the "pathway in the sky" picture in the A-6 VDI. Both aircraft have the same type of horizontal situation indicator. The F-14 has an inertial navigation system, but the INS was not reliable initially. Although the pilot's cockpit configuration makes the pilot's job awkward at times, the F-14's rear cockpit configuration is excellent. The RIO in the large rear cockpit has B-scope, E-scope, or modified PPI presentations for his radar picture. He also has the ECM controls and some weapon system switches duplicating the pilot's switches.

The Pacific Fleet fighter FRS, VF-124, was the first squadron to accept a Tomcat. VF-124 copied the F-4 FRS syllabus for the F-14's introduction, except for deletion of the F-4's air-to-ground work and addition of air-to-air gunnery flights. The Navy resurrected the designations VF-1 and VF-2 for the first operational F-14 squadrons, which made the airplane's initial deployment to the Western Pacific in the *Enterprise* in September 1974. The F-14 and F-4 shared the night and all-weather fighter role in carriers other than the *Essex* (27C) class until 1985. When the *Oriskany*, the last carrier of this type, decommissioned, two of the last night F-8 squadrons, VF-191 and VF-194, switched to F-4s in time to join the *Coral Sea* in her 1977 WestPac deployment. Those squadrons were disestablished during the 1980 reduction in carrier strength. VF-211, the last night Crusader squadron, converted from F-8s to F-14s and remains active. The number of F-14 squadrons increased until there were two Tomcat squadrons on all carriers except the two remaining *Midway*-class ships, the *Coral Sea* and *Midway*, which had two F-4 squadrons in their air wings.

．．．

When the Vietnam War ended, the Navy knew—and it has been confirmed—that the Soviets maintained a high respect for the carrier's offensive power. The Soviet navy had been increasing the number of missile equipped surface ships, submarines, and long-range bombers opposing the U.S. Navy's fast carrier task forces. During the 1970s and 1980s, the carrier navy's operations emphasized strike and fleet air defense exercises simulating a future, high intensity sea war against the ships and aircraft of the Soviet Union. Concurrently, the light and medium attack squadrons allocated about 20 percent of their training time to maintain proficiency in long-range, nuclear weapon strike tactics against enemy land installations. Interest in land and air wars, fought with conventional weapons, had ebbed again as it did after the Korean War.

While planning for a war at sea, it was important to the leaders of the U.S. Navy that no friendly or neutral ships be sunk at the beginning of a surprise war with the Soviets. It seemed extremely likely that a war could erupt without time to forewarn noncombatant ships to clear the war zone. One of the toughest challenges for the attack airplane crews was in differentiating between friendly, enemy, and neutral ships.[4] Because identification of ship targets was almost impossible with an A-7 or A-6A radar, attack aircraft sent to find and destroy the enemy would need to close within visual range of possibly hostile ships to identify them in daylight or darkness. That task would, of course, have been extremely hazardous in wartime. If a strange ship was hostile, the enemy could easily shoot down the attacker before the pilot could release his bombs.

War at Sea exercises practiced these procedures. A War at Sea operation began when the exercise umpire alerted the task force to the presence of simulated Soviet ships. Responding to the warning notice, the carrier would launch, day or night, all its strike aircraft in divisions on sector searches radiating from the task force's center. (Carrier aviators still used the same radial sector searches preferred in World War II and used by the AD and VC night units during the 1950s for locating and engaging an "enemy" fleet.) The search groups' mission was to locate, report enemy position, and strike immediately.

Once the search-and-attack groups identified the ships, the pilots had to convert to either a low- or a high-altitude bombing run while still maintaining positive contact with the target. That was extremely difficult at night for single-piloted A-7s.[5] For that reason, in the late 1970s, the A-7's principal role in the sea strikes was to provide SAM suppression sorties covering the air wing's A-6 bombers. The A-7s practiced staying at the edge of the Soviet ship's SAM envelope, preparing to fire Shrikes if their SAMs started tracking our aircraft.[6]

The A-6A squadrons were the primary bombers in the strikes planned against the Soviet naval units. For some time in the 1970s, the A-6 crews preferred low-altitude approaches to enemy ships to remain under the defense's SAM envelope. When within range, the attackers would use DIANE's toss-bombing mode to release their bombs. There were no opportunities to learn how many strike aircraft losses that tactic would have drawn during a war. As the Soviets improved their SAMs' accuracy against close-range, maneuvering targets, the low-altitude, toss-bombing approach lost favor. A high-altitude approach to a radar bombing run, while jinking to avoid SAMs, eventually became the preferred bombing maneuver for attacking Soviet ships at night or in adverse weather.[7] The fleet obviously had to have the potential to strike a Soviet surface force at night or in any weather as the enemy closed to within his missiles' range of friendly ships. However, few if any of the carrier air wings practiced night strikes against simulated enemy ships during the 1970s.

The A-6E's entry into the fleet during 1980 brought more-powerful engines, a new inertial system (CAINS), and an improved radar system. The A-6A radar operated with vacuum tube technology, but the new radar had digital circuitry, eliminating analog radar to digital computer interfaces. The new radar design eliminated the separate track radar, incorporating its function into one of three operating modes: search, target track, and AMTI. In the search mode, the B/N viewed a PPI-type presentation. When he switched to the target-tracking mode, the system displayed an expanded range scale view in a B-scope presentation. While in the tracking mode, the crew could switch quickly back to the radar's full wide-area search mode. Furthermore, the tracking mode was reliable. It gave the A-6E the bombing accuracy that was promised for the first A-6A but not normally achieved in operations. The more reliable CAINS seldom needed radar updates because of better gyros in the inertial platform and digital circuitry, reducing the system's drift rate. The CAINS, coupled with the Doppler navigation radar, increased the A-6E's accuracy in attacks on enemy ships or other targets.

About a year later, a modification to the A-6E added the Target Recognition Attack Multisensor (TRAM) system. In the A-6E TRAM airplanes, a central weapon system computer controlled three sensors: a forward-looking infrared system, a laser spot tracker system, and the radar. Soon after the Navy introduced the A-6E TRAM, the A-7Es began carrying a FLIR pod, which improved the pilots' ability to identify ships. The A-7E FLIR displayed its information on the airplane's primary flight information display, the HUD. It was difficult for an A-7E pilot to use the FLIR, because as he closed the range to a target, the FLIR sensor depressed when locked on the target. As the sensor depressed, the target's image on the HUD began to move down the glass display, presenting a confusing view to the pilot. Only an experienced A-7 night flyer could use his FLIR to full advantage. The advent of the infrared systems in the A-6E TRAM and A-7E FLIR aircraft did make it easier at night to separate Soviet warships from the large merchant ships common to all parts of the oceans. Pilots and B/Ns using the TRAM system could see details of ships as clearly as in daylight. Consequently, converting to an attack after identifying the enemy became easier also.

Precision and long-range standoff weapons became more prevalent during the 1980s, but few were effective at night or in wet weather. Laser-guided bombs (LGBs) were useful at night, but not in moist conditions. The A-6A had the Paveknife system for designating targets for the precision LGBs, but Paveknife was less accurate and harder to use than the LGB designator in the TRAM system. In addition, the A-6's system was modified to deliver the powered LGB, Skipper.

A few years after the TRAM version arrived, the A-6E units began carrying the Harpoon missile as the primary weapon for ship attacks. By 1982, the number of A-6 weapons and subsystems had increased to the point that A-6 crews carried a thick checklist manual for ensuring proper use of all weapons, particularly at night when switches and knobs were hard to locate in the cockpit.[8] The last weapon added to those used by the attack airplanes introduced prior to the Persian Gulf War was the Standoff Land Attack Missile (SLAM), using an infrared sensor with limitations similar to the LGB.

. . .

Task forces deploying in the 1980s continued the War at Sea exercises in which carrier groups, separated by hundreds of miles of ocean, "fought" against each other. In the early 1980s, the Soviets presented a powerful naval threat that was best engaged as far from our forces as possible. Air wing strike groups flew as far as 1,000 miles from their carriers in simulated attacks on simulated enemy naval forces in any weather. The carriers received information concerning the opposing force from national and Navy ocean surveillance systems. Consequently, the strike groups knew the general area in which to find their targets at the time the groups launched. The 1950s attack pilots who participated in similar strike exercises spent a good deal of their flight time searching for the location of the opposing force before they could attack; attack airplane pilots of the 1980s easily found their targets within their radar scan patterns when they first approached the enemy task force's expected position.

After the advent of the A-6E with an INS that did not need radar updates, the A-6s began to operate at low altitude in radar silence except to pop up for a final check of the target ship's position before firing a Harpoon. The E-2s could also provide final target location information passed from satellite sources. When the Harpoon was first introduced, the A-6 crews preferred to follow the Harpoons with bomb attacks, but

A-6E in flight showing TRAM turret on the forward bottom part of the airplane.
National Air and Space Museum

when the Soviets brought in the highly effective SA-N-6 SAM, follow-up attacks were no longer planned. The longer-range HARM replaced the Shrike as the antiradiation missile in the 1980s. Its appearance provided more standoff for the SAM suppressors covering the bombers in an attack.

By the mid-1980s, although the primary bomber was the A-6, the A-7s also carried part of the bombing load in day or night strikes, as well as acted as SAM suppressors. By that time, the A-6 and A-7E pilots were more proficient with the FLIR, making it possible for air wings to practice night attacks on opposing task forces.[9] At least one A-7E squadron, VA-83, practiced high-altitude "wheel" attacks, in which each bombing airplane in the strike group approached the threat's naval force on a different heading, spreading out the antiaircraft fire.

Pilots in VA-83's sister squadron, VA-81, trained at night to locate target ships by radar and FLIR and execute roll-ahead, low-angle dive

191

bombing attacks. At night, this meant concentrating on the HUD, FLIR, and radar displays while rolling inverted about 1,000 feet above the water. While low over the dark water in his approach to the target ship, the A-7 pilot waited until the target's radar image reached the point to commence the bombing maneuver. The night attack pilot then turned his attention to the flight instruments and began a constant-g pull-up to about a forty-five-degree pitch angle. He then commenced to roll inverted. (Obviously, engineers had improved the A-7E's attitude indicator relative to the A-4's, which was subject to precession errors.) At that point he pulled the nose down to a twenty- to thirty-degree dive angle, rolled upright, and searched for the target in the FLIR. After identifying the "enemy" in the FLIR's field of view, the pilot placed his aiming dot on the target and let the weapon system guide him to the proper bomb release point. Being able to execute that maneuver while maintaining 500 knots not far off the water at night required a skillful pilot.[10]

Along with the War at Sea practice strikes, the air wings in Task Force 60 flew actual surface search missions. That boring task common to World War II TBM crews had not disappeared, but the 1980s attack squadrons' job was slightly different. Their task was to locate and continuously shadow selected Soviet ships, with the goal of preventing a Soviet preemptive attack on the U.S. force. Remaining within weapon range of their assigned Soviet ship, A-6 or A-7 crews were ready night or day to attack if they saw the Soviets uncover their antiship missile batteries.

• • •

When John Lehman became secretary of the Navy in the early 1980s, he proclaimed that the Navy would follow a maritime strategy that pitted the fast carriers directly against Soviet Union forces in or near their homeland. To prove that a fast carrier task force could operate in close proximity to the U.S.S.R., the Second Fleet conducted exercises in the northern reaches of the Norwegian Sea, practicing strikes against Soviet facilities on the Kola Peninsula above the Arctic Circle. Cooperating with the Norwegians, the carrier task forces regularly steamed through the Norwegian Sea and operated for days within the fjords in northern Norway. That tactic served to screen the carriers from raiding Soviet bombers and attack submarines while providing sea room to conduct strike and air defense operations.

The *America* with CVW-8 participated in Exercise Northern Wedding during the summer of 1982. The carrier task group sailed to Norway's North Cape in almost complete electronic silence. The carriers turned off their radios, radars (including the air control systems), and TACAN in accordance with the exercise electronic warfare plan. That tactic kept the Soviets from locating the task force, but it also created quite a problem for the aviators. Nevertheless, even during long periods of electronic silence, the carriers conducted day and night flight operations. The A-7s and A-6s flew surface search missions, while the F-14s maintained CAP stations and intercepted several Soviet Bear aircraft that were searching for the task force.

The air wings established procedures for night and all-weather approaches to their carriers during the periods of electronic silence. Because there were no limits on the use of aircraft electronic systems beyond fifty miles from the carrier, these approaches used the airplanes' INSs in conjunction with the E-2's radar to establish a relative navigation system. While preparing for launch, all aircrews entered the carrier's current position into their airplane's computer. Flight leaders then entered rendezvous, mission waypoints, approach positions, and other mission position information relative to the carrier's initial position into the computer. The relative navigation grid established by those points became the reference for the air wing's planned departures from the carrier, en route navigation, and approaches to the carrier.

With some assistance from the E-2 controllers, who provided radar monitoring service during the departure and approach phases of a mission, the electronic silence procedure worked well most of the time. However, during Northern Wedding '82, at least one pilot did get lost. A pilot called on guard frequency, "I'm confessing, I'm climbing, I'm conserving, and I'm calling. If you don't want to lose a multimillion dollar airplane, somebody turn on a TACAN!"[11]

There were still accidents during relatively routine instrument approaches. One night the *John F. Kennedy* (CVA-67) steamed too close to a foreign shoreline to use the standard straight-in approach. The air wing had to use the alternate overhead, teardrop approach during one day and night of flight operations. The approach controllers and pilots had not practiced the teardrop for some time. Descending that night through a heavy cloud deck, an F-14 crew forgot the altitude requirements for the letdown and flew into the ocean. Despite the training, a change in all-weather procedures and a few seconds of carelessness in poor weather conditions caused fatalities.[12]

Carrier commanders conducted operational exercises in the North Atlantic under similar conditions to those experienced by the *Bennington* in 1953. Rough seas, clouds, and low visibility conditions still made carrier operations difficult, despite the improvements in systems and night and all-weather training. During those exercises, the Soviets probed the carrier task force's air defenses by sending long-range Bear bombers to locate the force and practice attacks on our ships. The fighters flew BAR-CAP, maintaining a barrier twenty-four hours a day between the carrier and the Cold War enemy.

It was common to spend over seven hours on one North Atlantic BAR-CAP mission in an F-14, waiting to intercept Soviet Bears searching for the task force. Maintaining the fighter barrier required KA-6 tankers to constantly refuel the F-14s. The fighter commanders perfected a tactic, called Chainsaw, to intercept the Bears at maximum range from the task force center. KA-6 tankers would team with an F-14 for multiple refuelings as the fighter flew out to meet the Bears. By combining early detection of the Bears with the long range of the F-14s when refueled in the air, the F-14s made their intercepts hundreds of miles from the task force.

Although ship or E-2 controllers set up the fighters for the intercepts using information about Soviet bomber movements obtained from satellite detection systems, the fighters had complete control of the final phase of the long-range intercepts, just as Henry and Tucker did to finish their successful night engagements in earlier wars. The F-14 crews drilled in forward sector approaches for a Phoenix or Sparrow shot, followed by a stern conversion for a Sidewinder or gun firing opportunity. Procedures using the same Punch-and-Judy protocol between an E-2 controller and a fighter crew had changed little since World War II. Although night intercepts were routine, there was a difference. One F-14 fighter pilot explained the difference between day and night operations rather simply: "You can't see s—!"

Exercises conducted by Task Force 60 during the winter of 1984 complemented the Second Fleet operations. The fighter crews planned and executed maximum-range night or day intercepts similar to the drills they flew in the North Atlantic. They planned to use the Phoenix as their primary weapon, firing beyond visual range without positive target iden-

194

tification. However, just as in the North Atlantic, for training purposes the Sixth Fleet F-14 pilots closed approaching Soviet Bears and Badgers to within gun range.[13]

The Pacific Fleet carriers executed the 1980s maritime strategy by practicing strikes against the Soviet facilities on the Kamchatka Peninsula in Siberia. That strategy required long-range A-6 and A-7 attacks against airfields and naval bases while protecting the fleet against missile-carrying Soviet bombers. There were extended exercises in which multicarrier task forces continued flight operations continuously for several days.

. . .

In past years the Navy had used many fighters as bombers. The list of fighter-bombers included the Hellcat, Corsair, Panther, Banshee, and Phantom. Furthermore, all but the Panther had accomplished those tasks at night. The F/A-18 Hornet became the next in that distinguished series of night fighter-bombers and the first airplane designated as a fighter/attack airplane.

The F/A-18A appeared in the fleet in 1980, but had several modifications early in its lifetime that soon changed its designation to the F/A-18C. The F/A-18C is a single-seat, twin-engine airplane about 20 percent smaller than the F-14. It is the heaviest single-seat airplane the Navy has owned. The Hornet's top speed is near Mach 1.5, somewhat less than the F-4 or F-14, but easily fast enough to match any airplane in the world in combat. It quickly turns its extra energy to acceleration. That characteristic makes the difference in a fight. Because it was designed as a small airplane, the Hornet has a short range and short endurance. In open-ocean operations, the fuel reserve requirement for the airplane, if not refueled, forces carriers' cycle time down to one hour and fifteen minutes rather than the one hour and forty-five minutes that most carriers used for routine operations during the early 1980s. The Hornet carries Sparrow, Sidewinder, and the new AIM-120 air-to-air missiles. It has a rapid-fire 20-millimeter cannon, and also carries precision and general-purpose air-to-ground weapons.

Computers control all systems in the F/A-18. Furthermore, engineers can adjust the computer's flight software to change the flying qualities of the airplane. The test pilot who flew the F/A-18 on its first flight said it flew just like its development flight simulator. There was nothing myste-

195

rious about its handling characteristics. The F/A-18's flight system software provides greatly improved control when following intercept or ACLS data link commands. A HUD is the principal instrument for visual and instrument flight. Because the HUD shows the pilot all the basic flight information he needs, he does not need to switch from visual to instrument scans. Computers drive the HUD and three Digital Display Indicators (DDIs) that present all flight and engine information to the pilot. Six-inch scopes offering horizontal and vertical situation information are the main presentations.

The pilot can switch quickly between the air-to-air and air-to-ground weapon modes of the radar. The radar can lock on and automatically maintain track air targets. For air-to-air engagements, the F/A-18's Doppler radar gives it a variable vertical search. For air-to-ground missions, the

The F/A-18C instrument panel. Note the HUD—the principal flight instrument—and the three DDIs.
Boeing Company

196

radar has an AMTI mode. Using the Doppler beam-sharpening radar feature, the pilot can lock on and track ground targets. An improved FLIR makes performance of air-to-ground weapon tasks at night easier than in the A-7E, but the F/A-18C's FLIR does not have the wide field of view, variable range system, or resolution of the A-6E FLIR. The F/A-18C has the capability to carry and deliver LGBs on targets designated by A-6s.

VA-125, in earlier years the Pacific Fleet's A-4 FRS, was redesignated VFA-125 and became the first F/A-18 fleet replacement squadron in November 1980. The first commanding officer was Cdr. John Lockard. When the F/A-18 FRS was established, the instructors divided the F/A-18 syllabus into fighter and attack phases. The fighter phase was almost identical to the F-14 syllabus. It ended with a deployment to Key West, Yuma (Arizona), El Centro, or Fallon (Nevada) to use a Tactical Air Combat Training System (TACTS) range. The attack phase followed a course identical to the A-7 FRS syllabus. The F/A-18 attack course finished with a deployment to one of the same outlying air stations for air-to-ground weapons training.

Two A-7 squadrons were redesignated VFA-25 and VFA-113 and, with two newly established units, VFA-131 and VFA-132, became the first Navy F/A-18 squadrons. About half the new squadrons' carefully selected pilots had fighter backgrounds and the other half had attack experience. The pilots with appropriate backgrounds took responsibility for cross-training their mates in their respective specialty. Unlike in some early dual-mission Banshee squadrons, cross-training soon gave all pilots experience in both fighter and attack tasks.

The F/A-18's first carrier deployment came when VFA-25 and VFA-113 went to sea in the *Constellation* in February 1985. VFA-131 and VFA-132 joined two Marine Corps squadrons in the *Coral Sea* during August 1985.[14] F/A-18s soon replaced the F-4s and A-7s in the *Midway*'s air wings. Then the F/A-18s began replacing the A-7s in the rest of the Navy's light attack squadrons.

• • •

Carrier task forces faced enemies again within a decade after the Vietnam War. The first of several incidents requiring carrier air power occurred when Iranians kidnapped about fifty U.S. citizens in Tehran in 1979. That situation prompted President Carter to send and then keep at least one aircraft carrier on station in Southwest Asian waters for an indeterminate time period. Along with several other Pacific Fleet CVs, during 1980 and 1981 the *Midway* and its air wing deployed five times to the Indian Ocean from its home port in Japan.

The carriers and air wings in the Indian Ocean emphasized night operations to be ready for a quick response to a presidential order to strike. Fighter, light attack, and medium attack squadrons were equally ready to perform their night missions. In the Indian Ocean, carriers commonly operated without a divert field in case an emergency closed the carrier's flight deck. Consequently, the commanders required the airplanes to have a large fuel reserve when returning to marshal, particularly at night or in adverse weather, after a training mission.

In conditions similar to those in which the 1950s WestPac flight operations were conducted, the air wings used a spar towed by the carrier for visual and radar bombing practice, both day and night.[15] Attack squadrons' night missions included surface search, simulated night bombing, and aerial refueling support. By that time, the A-6 squadrons' KA-6s were the air wings' primary tankers, although the A-7s and S-3s took some of the refueling missions.[16]

In 1981, Libya's Moammar Gadhafi began a series of actions blatantly defying the Western nations. He declared that the Gulf of Sidra belonged to him, defining the dividing line between the Mediterranean Sea and the gulf as the "line of death." The president objected to Gadhafi's delineation of his sovereignty and ordered the Navy to exercise the ships of Task Force 60 in the Gulf of Sidra.

Exercise Burning Wind took place off Libya in early 1981. The two carriers in Task Force 60, of which one was the *John F. Kennedy*, operated on a twelve-hour flight schedule, 1000 to 2200. The *Kennedy* air wing's attack squadrons flew day and night surface search and attack missions. Although the Libyans did not send any of their missile boats against the ships of the task force, the air wing had a plan for attacks against enemy warships. The plan included use of the A-7s as Shrike shooters and the A-6s as bombers or Standard ARM shooters. The F-14s would maintain CAP in defense of the task force and provide air cover for the strike.[17] The climax of the 1981 Exercise Burning Wind occurred with a day engagement between F-14s and MiGs that ended with the score, Navy 2, Libya 0. There were no fighter engagements at night during that first confrontation with the Libyans.

Task Force 60 continued exercises in the Gulf of Sidra for several years after the 1981 incident. In March 1986, during the exercise "Operations in the Vicinity of Libya III," TF-60, with the *Coral Sea, America,* and *Saratoga,* again confronted Gadhafi's forces. The carriers continued to alternate in twelve-hour shifts, as Task Force 77 carriers had done in the Vietnam War. Airplanes were airborne at all hours, while maintaining a reasonable carrier crew rest schedule. The bigger decks, the *America* and *Saratoga,* shared the night duty.

On the night of 24 March, the airborne E-2C detected two surface contacts approaching the task force. A section of VA-34's A-6s investigated the contact and identified them, using their FLIR sensors, as Libyan missile boats. The task force's rules of engagement permitted destruction of Libyan craft approaching within weapon firing range of any U.S. warship. When the hostile boats reached the critical position, one A-6 fired a Harpoon antiship missile. The pilot of an escorting HARM-equipped A-7E saw an explosion light up the night as the Harpoon hit. After losing radar contact with the boats, but guided by vectors from the E-2, the A-6 crew pressed the attack to drop CBUs. The second A-6, still maintaining radar contact with the enemy, fired its Harpoon. The first A-6 was almost over the enemy boats on its CBU run when they saw the second Harpoon explode on the target. That same night the airborne A-7 SAM suppressors hit an active SA-5 SAM site on the beach that was within range of American airplanes.

. . .

Later that spring, terrorists bombed a Berlin nightclub, an act threatening the United States and its European allies. President Reagan authorized reprisal strikes composed of Air Force and TF-60 aircraft against Libya, the alleged instigator of the nightclub attack. Nicknamed El Dorado Canyon, the operation was an attempt to destroy Gadhafi's headquarters. In the early morning hours of 15 April 1986, Air Force F-111s launched from England and hit Gadhafi's headquarters in Tripoli and Sidi Bilal airfield. The carrier air wings in the *Coral Sea* and *America* attacked the army barracks in downtown Benghazi and an airfield, Benina, on the outskirts of Benghazi.

This was the first naval night strike in a heavily defended area since 1973, thirteen years before. Some Naval Aviators with Vietnam War experience under their belts were in senior billets. Commander Lockard was now the *Coral Sea*'s executive officer. Rear Adm. Jerry Breast, who had flown A-4s in Vietnam, was commander, Task Force 60. The only aviators with combat experience in the air wings were now squadron commanding officers.

The night of the attacks, the Intruder, Hornet, and Corsair II ready rooms were quiet. Affected by the unknown dangers of combat, the Intruder aircrews were also concerned because they had had little practice acquiring and attacking land targets with radar at night. The attack airplane crews had spent most of their flight time at sea "rigging" ships, using FLIR to identify ship targets.

Just as in the Gulf of Tonkin, the carriers conducted flight operations with full flight deck floodlights, but they did reduce electronic signals during the strikes. Fourteen A-6s, six A-7Es, and six F/A-18s from two carriers composed the carriers' strike groups, which launched in time to arrive at the planned target time, 0200.[18] Two A-6s from VA-55 aborted after takeoff because of weapon system failure.

Although it was known that the Libyan fighter pilots were not trained to fight at night, night fighters filled CAP stations around the carriers. As expected, there was no Libyan night fighter opposition. The light attack aircraft provided SAM suppression for the A-6s. During the short period of the strike, the SAM suppressors expended about 95 percent of the Sixth Fleet's HARM supply.

It was not very different over Benghazi that night than it was over Vietnam, or for that matter in Taipei harbor with Bill Martin's TBMs forty years before. VA-55's six Intruders proceeded at low altitude through SA-3, SA-6, and antiaircraft gun defenses to their airfield target. As the night's fireworks display from air defense weapons raised the crews' adrenaline, the A-6s strung out in a trail with enough interval to avoid the preceding aircraft's bomb blasts. The Libyans obligingly left the Benina airfield's runway lights on, so the aircraft achieved amazing results for their night's work. Using the same tactics, VA-34's B/Ns picked out and hit the Libyan army barracks with little or no damage to nearby apartment buildings. All twelve aircraft escaped unscathed.

■ ■ ■

Between 1986 and 1989, the fleet commanders gradually became less concerned about the Soviet bomber and ship missile threats. In the North Atlantic, Mediterranean Sea, Sea of Japan, and Northern Pacific waters, there were fewer Soviet Bear or Badger bombers coming out to investigate the carrier task forces' activities. By 1989, intelligence sources confirmed that the Soviets were not the powerful threat to U.S. naval forces that they once were.

During the summer of 1989, the Soviet Union began to split, as the basic economic structure of the U.S.S.R. showed its flaws. In the midst of the upheaval and international realignment happening that year, the remaining plausible threat to the United States came from small, relatively weak nations bent on tweaking the eagle's tail. U.S. defense policy and military strategy had to change.

A change in the national concern about human life also became more apparent in the 1980s. Public opinion was that, in any future war, the military should not accept *any* casualties. Consequently, the military, including Naval Aviation, began changing its tactics to achieve the desired goal of eliminating friendly losses in any future war, even at the expense of combat effectiveness. Fighter and attack tactics had to change to fit a new military strategy and to accommodate the perception that wars can be fought without casualties.

Predicting that combat losses at night would be less than in daylight, in the late 1980s the fleet replacement and combat squadrons began more extensive night training. (Even before the breakup of the Soviet Union in 1989, light and medium attack squadrons practiced night bombing to maintain proficiency in preparation for a "third world" war.) The proportion of night flight time, about 25 to 30 percent, did not change much in the squadrons, but the intensity of night training increased. More flight time was dedicated to night formation flying, as that skill is essential for night group strikes. The squadrons' training programs also added more flights devoted to basic night dive bombing and night intercept training rather than simple night navigation flights. Some air wings practiced night coordinated strikes on the Fallon ranges using the FLIR as the principal sensor for both the A-7s and the A-6s. Although the squadrons preferred to practice night weapon deliveries on shore range targets, on the *Saratoga*'s deployment in 1987, VA-83 used smoke lights and a towed spar as targets to keep some degree of night bombing proficiency while at sea. A-6Es practiced using laser-guided bombs at night as well as in the daylight.[19] Interest in the Soviet naval threat waned, but air wings planned for coordinated night attacks against third world fast missile boats. That threat, first appearing in the Vietnam War, still confronted the Navy.

．．．

When the A-6 and F-14 became the primary combat airplanes in the air wings, the pilots had flight instruments in their cockpits that almost eliminated the need for the basic instruments of previous carrier airplanes. The maneuvers that Hunter and Zacharias managed in the Vietnam War, as well as the night roll-ahead bombing practiced by A-7s, were possible because of the confidence that pilots had in the modern class of flight instruments for night and all-weather flying. By the time the Cold War ended, equipment in the night airplanes made night and all-weather flying relatively simple—if the systems worked.

All combat airplanes had computer-driven, digital primary instrumentation systems. Moreover, each airplane type had redundant displays for primary flight information in case system components began to fail. As an example, the F/A-18's flight instrumentation system had four modes. Even at night and in full instrument conditions, the head-up display contained all primary flight information and was the focus of the pilot's attention. If that failed, the pilot's scan moved to the DDIs in the center of the instrument panel, just below the HUD, where the primary flight information could also be viewed. At the third level of degradation, the pilot reverted to digital attitude, heading, and navigation indicators, but used old analog airspeed indicators and altimeters. The last resort, if all the computer components and displays failed, was the collection of 1960s analog airspeed, altitude, turn-and-bank, rate of climb, and attitude instruments. These were placed below the electronic displays in the instrument panels. The carrier pilot could not forget how to fly using information acquired from analog instruments in case of computer or electrical system failure but had little or no practice in using those basic instruments.

Weapon system technology developments also improved the effectiveness of all the night combat aircrews. FLIRs on F/A-18s as well as A-6s greatly improved the ability of the night pilots to locate and hit targets. Radar bombing using the Doppler beam-sharpening feature of the F/A-18's radar became standard for night bombing in that airplane. The new standoff weapons promised elimination of aircraft losses, but most of the new weapons were not usable at night or in bad weather. The

SLAM showed promise as a night weapon because it had an infrared sensor. However, all such sensors, including FLIRs, lost effectiveness in rainy, cloudy, foggy, hazy, or smoky conditions. LGBs suffered similar limitations. Moreover, radars also had some weather limitations, particularly in heavy rain. By early 1990, Marine Corps pilots were using the new night vision goggles (NVG). Use of the NVGs made possible effective visual detection of small tactical targets such as tanks or trucks on the darkest of nights, thereby increasing the effectiveness of single-seat airplanes conducting night interdiction and troop support missions.

Because the nature of a future threat was uncertain, under normal circumstances, our night and all-weather fighters could no longer risk firing missiles in a BVR situation. The pilots had to obtain positive identification of a possible enemy before firing. The F-14 and F/A-18 squadrons reduced the emphasis on firing air-to-air missiles at targets beyond visual range, making short range air-to-air missiles and cannon necessary again. Night fighter pilots renewed interest in closing a bogie for stern conversion intercepts. Fighter pilots practiced keeping position as close escort for a strike group at night or providing an airborne defense over a land target complex. Furthermore, in some air wings, fighter squadrons planned coordinated night fighter sweeps into enemy territory to destroy their airplanes on the ground and in the air.

The FRS's instructor pilots challenged the student fighter and attack pilots learning the rudiments of fighter and defensive tactics on the TACTS ranges. The TACTS range instrumentation provides position and timing information that allows the participants in practice coordinated strikes to see how well they actually execute the strike. The value of that information cannot be overestimated. As the tracking accuracy of the ranges improved, the Strike University and Top Gun schools experimented with night air combat maneuvering on the instrumented ranges. By 1989, the schools and fleet squadrons used the TACTS ranges for training multiplane night strike groups. The night training required wide altitude and distance separation between airplanes in the strike groups. There were few, if any, aggressor fighters. If aggressors were used, the exercise rules specified wider separation between aircraft for safety purposes.[20]

Chapter 9

THE PERSIAN GULF WAR

During August 1990, Saddam Hussein's Iraqi troops roared across Kuwait in a little less than two days, stopping unopposed at the Saudi Arabian border and thus beginning the Persian Gulf War. The United States honored its commitment to protect Kuwait and Saudi Arabia by forming a coalition with the most powerful European nations. That coalition promised to force Iraq out of Kuwait and restore its old borders.

After deploying about a half-million men and 1,800 combat airplanes, on 17 January 1991, the Coalition struck Iraq with overwhelming air attacks. Ready again to demonstrate their night and all-weather potential, the *Midway* and *Ranger* were in the Persian Gulf. The *Saratoga*, *John F. Kennedy*, and *America* were in the northern part of the Red Sea, and the *Theodore Roosevelt* (CVN-71) was en route to the Persian Gulf. There were ten F-14, nine A-6, nine F/A-18, and two A-7 squadrons in the six carriers' air wings.

At 0238 in the early morning darkness, the carrier air wings joined Air Force bombers and combat support aircraft in the first raids on principal Iraqi military installations. During the first night, the *Ranger*'s and *Midway*'s air wings struck airfields in the complex west of Basra. The *Ranger*'s air wing also flew mining and the first of many surface search and attack missions to destroy Iraqi warships. The air wings of the *Saratoga*, *Kennedy*, and *America* hit the Iraqi airfields in the western part of the country nearest the Israeli and Jordanian borders.

From the aircrews' point of view, the Coalition air planners did an outstanding job coordinating all of the strikes the first night. Airspace

A VF-41 F-14 returns from a predawn CAP over Iraq. *Lt. G. B. Parsons, Department of Defense*

management by the E-3 AWACS (Airborne Warning and Control System) and E-2 controllers was excellent, maintaining a high degree of coordination between strike groups throughout the war. "Red Crown," the sea-based air defense coordinators in the Red Sea and the Persian Gulf, assisted the airborne controllers directing the air defense of the Coalition forces.[1]

All the carriers had two F-14 squadrons for day and night fleet air defense missions except the *Midway*, which had three F/A-18 squadrons. The first night, the *America* provided night fighters protecting the Red Sea task group, while the *Ranger*'s F-14s had the same duty over the Persian Gulf and Arabian Sea. The F-14s continued to fly defensive missions and maintain airplanes in alert status twenty-four hours a day, ready to counter Iraqi air attacks on the task forces, for the first three weeks of the war. Strike groups in the air wings with only F-14s had few fighter escorts

205

over Iraq for that period. After the initial focus on fleet air defense, the F-14 crews and some of the F/A-18 pilots flew defensive missions night and day in the vicinity of the Iraqi installations or they flew "attached CAP," staying within sight of the strike flight they were assigned to protect, ready to intercept any threat to the attack aircraft. The F-14s and some F/A-18s also flew fighter sweeps, roving north toward Baghdad.[2]

The Iraqi air threat was almost nil, but a few air-to-air engagements did occur during the first days of the war. Although the airspace controllers in E-2s and Air Force AWACS warned the strikes about airborne MiGs, a MiG-25 probably shot down a *Saratoga* F/A-18 in an engagement on the war's first night. That Hornet was alone on a SAM suppression mission in the vicinity of known MiG activity when it disappeared. Because there was no known SAM activity in the area, the MiG received credit for the F/A-18's destruction.

During the first day's operations, Lt. Cdr. Mark Fox of VFA-81 in the *Saratoga* shot down a MiG while on a strike mission. Although the engagement occurred in daylight, the tactics his flight employed were typical of an F/A-18 or F-14 night air-to-air engagement. The mission of Fox's strike element was to bomb aircraft and hangars on the "H-3" airfield in the western part of Iraq.

The flight passed the border in a loose combat spread formation, Fox using his radar to search the air below and forward of the formation. He heard calls from the E-2 controller warning of strikes of airborne MiGs; however, he was not sure of the position of the MiGs because of the number of "Bullseye" calls. He did understand that the E-2 controller had declared an aircraft over H-3 as hostile. Suddenly, the controller called on his strike's tactical frequency, alerting them to MiGs that were fifteen miles directly ahead of their formation. With his armament system already set for releasing bombs, Fox immediately switched to the air-to-air mode and picked up a radar blip approaching from H-3. The blip had to be the Iraqi MiG because there was no IFF showing and it was closing on his flight with a relative speed near the speed of sound, over 1,200 knots.

Fox reacted in a second, listening for a Sidewinder tone indicating that his missiles were locked on, but he couldn't get a good missile tone. He kept caging the Sidewinder seeker to look down the radar line-of-sight directly at the oncoming MiG. Fox heard the Sidewinder's distinc-

Lt. Cdr. Mark Fox on the *Saratoga* during the Persian Gulf War. *Department of Defense*

live lock-on growl at the same time he saw the speck of the Iraqi fighter in the distance, and he squeezed the trigger. The missile fired with the roar of a passing train, then simply just vanished. Their relative speed rapidly brought the MiG closer and bigger in Fox's eyes. Assuming the Sidewinder wasn't working, he thought as he selected a Sparrow that the MiG would not get away from the larger missile and squeezed the trigger again. This time he could clearly see the missile streak toward his target. As the Sparrow accelerated, the enemy briefly disappeared in a bright flash and cloud of black smoke—the Sidewinder hit!—then emerged, still nose on but trailing flame and smoke. A moment later the Sparrow hit the doomed fighter with another explosion. Incredibly, there was still an airplane there, albeit clearly burning, decelerating, and descending as it passed through Fox's section.[3]

At night, fighter pilots could duplicate Fox's air-to-air engagement, but only with authority to fire beyond visual range. If required to get a visual identification of a possible enemy aircraft at night, a fighter pilot facing a rapidly closing bogie would try to perform a stern conversion for visual identification of the threat and a Sparrow, Sidewinder, or gun

207

shot. At the speed of such an engagement, it is unlikely that the fighter pilot would be successful in getting a firing opportunity.

The Navy and Air Force SAM suppression tactics on the first night and throughout the succeeding days were outstanding. EA-6B Prowlers, besides being HARM shooters, provided jamming and passive ECM for the Navy strikes and some of the Air Force strike groups. The F/A-18s and A-7s not only used HARM but also launched decoys resembling aircraft to fool Iraqi SAM operators (the Israelis had first used SAM decoys successfully during their strikes against the Arabs during the 1980s). Prior to launching from the *Saratoga*, F/A-18 pilots assigned to suppress SAMs in the Baghdad area received information about the specific SAM sites that each was to cover. The night strike pilots reported observing the HARM warhead explosions impacting on or near the locations of SAM radar sites.[4] The decoys, HARM, and jamming plus passive ECM tactics effectively neutralized the Iraqi SAM network during the war.

Strike tactics quickly changed in this war (see figure 5). The first night, the A-6s used low-altitude flight profiles to get under the SAM defenses and deliver their loads of 500-pound bombs, Rockeye cluster bombs, and mines. The A-6 strike pilots learned again that heavy barrage antiaircraft gunfire was deadly against airplanes at low altitude, even at night. Barrage fire shot down one of VA-35's A-6s from the *Saratoga* and heavily damaged another during their first strike on an airfield in western Iraq. Both A-6s were at low altitude. A-6s from the *Ranger* struck an airfield target near Basra. They also faced intense barrage fire at low altitude, but were lucky; their airplanes received no damage. However, during the necessarily low-altitude mining missions conducted by the *Ranger's* A-6s, one VA-155 airplane and crew were lost. The flag officers in command during the Persian Gulf War understood that the American public did not want any losses. As a result, the air commanders sacrificed weapon accuracy to increase the aircrews' survivability. After the first night, all attack and fighter-attack squadrons released their weapons from 10,000 to 20,000 feet above the surface, avoiding most antiaircraft gunfire and the newer handheld infrared-guided missiles.[5]

. . .

The first phase of the air campaign continued until the end of January 1991. The carriers settled into a routine flying schedule, striking Iraq's primary military and industrial targets. In the Red Sea, the three carriers

Antiaircraft fire in Baghdad on the first night of the Persian Gulf War. *Copyright 1997 by Cable News Network, Inc. All rights reserved.*

rotated on a dawn to dusk, noon to midnight, and midnight to noon schedule. The carriers did not operate on a land-and-launch cycle, but launched about twenty-four airplanes during each day and a group of about twenty airplanes each night. The attack airplanes split into four to six strike groups to hit different targets or aimpoints on the same target.

Time in the air for the strike groups flying from the Red Sea to the Baghdad area averaged 4.6 hours because the strike airplanes had to refuel several times en route to and returning from their targets. A night combat sortie was apt to be boring for the Red Sea aircrews, who spent hours strapped to their seats watching nothing but the instruments, the dark sky, and a few stars. Each carrier in the Red Sea had a ready deck when strikes returned from their missions. At night or in the marginal weather conditions existing in the Middle East that year, the night refuelings and carrier landings were the most challenging parts of the strike missions.

The carriers carried a limited number of precision standoff weapons in their magazines. These "smart" weapons were used in preference to

209

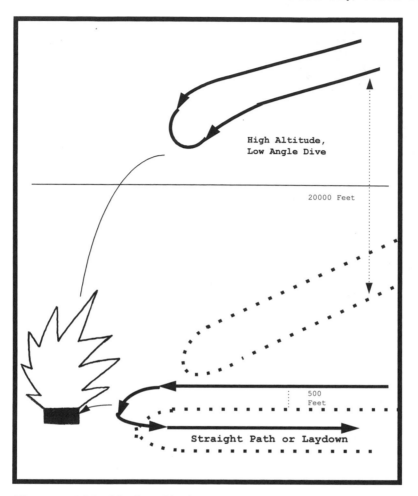

Figure 5. High-altitude and laydown tactics during the Persian Gulf War.

general-purpose bombs when they were available. The A-6Es with TRAM were the only carrier airplanes with a laser designator that enabled a single airplane to use the LGB. Consequently, the A-6s delivered most of the LGBs during the war. The Mk-80-series general-purpose bombs were still the bread-and-butter air-to-ground weapon, despite the desirability of the standoff weapons.[6]

As always, the weather affected the air campaign. Clouds or other moisture in the air prevented the use of available LGBs and reduced the effectiveness of FLIRs. By coincidence, during the period of the air campaign, the weather over Iraq had more clouds and moisture than usual. Although the FLIR had become the preferred night attack sensor for A-6s and F/A-18s, weather conditions and operational rules (staying above 10,000 feet) limited about half of the night strike missions to radar-aimed weapon deliveries. Because the A-7 FLIR's effectiveness was still poor, the night bombing A-7s relied on radar or eyes alone in all conditions throughout the war.

After the first night's strikes, the F/A-18s joined the A-6s in the night attacks on assigned targets. Because an airfield or an industrial facility had so many aimpoints, often the whole night launch from the *Saratoga* would hit one target complex. Both the A-6 and the F/A-18 groups, after flying the four hundred miles to the Iraqi border, penetrated Iraqi territory as individual airplanes at about 20,000 feet. Following the EA-6B and F/A-18 SAM suppressors shooting HARMs, the strike's A-6s were the first into the target area.

An F/A-18 from VFA-136 in flight over the *Dwight D. Eisenhower* (CVN-69) in the Persian Gulf. *CWO2 Tony Alleyne, Department of Defense*

211

For the first few nights, the F/A-18s followed the A-6s by one or two minutes, but learned that the A-6s stirred up the antiaircraft fire. Large-caliber bursts and some of the 23-millimeter tracer streaks appeared to be reaching the F/A-18's release altitude. At night, it was easy to see the steady stream of the rapid-fire 23-millimeter guns, just as it had been easy to see the same effect during the Vietnam War. Adapting to the situation, the F/A-18s began separating from the A-6s by seven minutes. The *Saratoga's* F/A-18s timed their departure so that they could get the desired difference in time-on-target but still use their faster cruise speed. At cruising altitude, there was about a 100-knot difference in the best cruise speed for the A-6 and the F/A-18, but the Hornets lost some time because they slowed to refuel more than the A-6s. Consequently, over the long distances flown by the strike groups, the F/A-18s delayed departure from the carriers for ten to fifteen minutes after the A-6s. Within the longer time interval between strike groups, the Iraqi gunners seemed to relax, letting both the A-6 and the F/A-18 groups through the target area without facing heavy antiaircraft gunfire. After the Iraqi air defenses had been beaten down, in the latter stages of the air campaign's first phase, attack groups separated by only two- to five-minute intervals could safely follow each other over a target.[7]

The Persian Gulf carriers operated off the coast of Bahrain, about three hundred miles from the Basra target area, for about two weeks after the air campaign opened. As soon as the admiral recognized that there was no air threat to the carriers in the Persian Gulf, he moved the operating area north to about 150 miles from Basra. From that launch position, the carrier land-and-launch cycle shortened from two hours to one and a half hours, increasing the number of sorties by about 30 percent. The *Ranger*, after operating for an unbroken forty-eight hours at the beginning of the war, became the Persian Gulf night carrier, its airplanes flying from dusk until dawn. Her air wing with A-6 and F-14 squadrons was well suited for night work. The *Midway* and *Theodore Roosevelt* split the twenty-four-hour flying day, with one CV flying twelve hours while the other rested.

The squadrons in the Persian Gulf carriers used night strike tactics similar to those of the units flying from the Red Sea. A typical strike on a large target complex occurred near the end of the first phase of the air campaign. Four *Ranger* A-6s executed a night strike on the Al-Basra

A-6 and F-14 on the *Ranger*, the Persian Gulf War's night carrier. *U.S. Navy, courtesy Cdr. S. B. Barnes*

power plant with each A-6 carrying five 2,000-pound bombs. Lt. Cdr. "Boots" Barnes, the strike leader, had planned the strike so that the A-6s would arrive at their individual bomb release points thirty seconds apart.

The A-6 group launched from the *Ranger* about 2200 and separately penetrated Iraqi airspace above 15,000 feet, keeping 4,000 feet vertical separation between airplanes. The individual A-6s relied on their INSs to provide the precision required for accurate time separation. Each A-6 crossed timing positions while heading in different directions toward the target. The power plant complex with its five smokestacks, large generator, and boiler buildings made an excellent FLIR and radar target. The B/Ns saw the target's radar return from about fifty miles away, but waited about a minute and a half for the power plant's FLIR image to appear.

Barnes had each A-6 hit a different aimpoint in the complex. As each bomber reached a point about ten miles from its aimpoint, the pilot pushed over into a twenty-degree dive. Bombardiers kept their aimpoints in the FLIR aiming circle and let their weapon system computer release the bombs at the correct instant. After each airplane's bombs released at 10,000 feet, the night strike pilots, concentrating on their "pathway in the sky," pulled up and to the right, each A-6 wheeling off the target toward home. At that altitude, the antiaircraft fire was not a threat. Furthermore, Barnes said that he did not see any SAMs tracking on the strike, presumably because of the effective active and passive SAM suppression provided by the air wing's EA-6Bs.[8]

. . .

As January ended, the Coalition began the air campaign's second phase, battlefield air interdiction (BAI). During this phase of the campaign, the daily air planners assigned strike groups to "kill boxes," attempting to destroy the Iraqi ground troops entrenched in Kuwait and southern Iraq. Each kill box covered an area about thirty miles on each side. The three carriers in the Red Sea flew nearly seven hundred miles one-way on BAI missions. The *America* finally moved from the Red Sea into the Arabian Sea on 15 February to provide more carrier air sorties and firepower over the kill boxes.

The A-6s from the Persian Gulf carriers routinely carried two LGBs plus twelve 500-pound bombs or Rockeyes on the night BAI missions. The Red Sea A-6s carried external fuel tanks and fewer bombs. All F/A-18s and A-7s carried up to six 500-pounders. The Coalition's daily flight plan directed strike groups of four to six attack airplanes to be at a timing control point (TCP) at a scheduled time. When strikes reported at their TCP, the ABCCC or an E-2 took control of each strike flight. Because of fuel limitations, the Red Sea carrier strike groups received priority to enter their kill boxes. The groups from the Persian Gulf could afford to linger before receiving clearance to enter their assigned kill box.

Sometimes the kill box controllers gave specific aimpoint coordinates within the box to a flight. More often, however, the night strike groups would spread out, remaining at 15,000 to 20,000 feet, with each attacker taking slightly different headings and time separation over the thirty-mile width of the group's kill box. When the weather was clear over the flat terrain of the Iraqi desert, A-6 crews and F/A-18 strike pilots using FLIR could easily see the tanks, bunkers, trucks, and troops of the enemy. If a pilot detected a hot spot on the FLIR, or a blip on the radar if in an A-7, he pushed over in a shallow dive to release an LGB or bombs. If there were secondary fires, the pilot transmitted the target coordinates to the other members of the flight, who joined the attack, just as was done on night interdiction missions during the Korean and Vietnam Wars. The A-6s released their second LGB with several minutes between bombs. That interval allowed the dust to settle so the second LGB could get a better hit.

On nights when the weather prevented use of the FLIR or laser, the A-6s depended on the radar's tracking or AMTI mode to make a successful bombing run. The F/A-18's Doppler beam-sharpening radar mode gave that airplane similar accuracy as the A-6. The night strike pilots reported reasonable hits even with general-purpose bombs dropped from release altitudes over 10,000 feet. Without FLIR or a high-resolution radar, pilots would have had no chance of seeing, much less hitting, trucks or tanks from 10,000 feet at night. After passing through the kill box, crews would reset the weapon system computer for new coordinates and make another run. The passes through the area continued until each airplane expended all weapons or the group's time in the target area ended.

Air planners concentrated on the kill box labeled the "Road to Hell" as the air campaign drew to a close. As Coalition air power hammered the Iraqis, their troops tried to leave Kuwait using tanks, trucks, or any other vehicle that would run. The objective of all night strikes was to stop or destroy those vehicles. Orbiting over the Road to Hell, the A-6 B/N or light attack pilot used his FLIR to scan the terrain looking for tanks, truck parks, or moving vehicles, releasing LGBs or bombs on visible targets. Sometimes the gunfire from the ground was colorful, but not menacing, because by that time in the air campaign, antiaircraft fire was severely reduced. The attack aircraft began using a 6,000- to 7,000-foot release altitude for their weapons.

BDA immediately after bomb impact continued to be a problem, particularly at night. Even with FLIR, it was difficult to judge where bombs hit or to estimate damage to targets. In one example of poor BDA, a crew reported in their immediate flight debrief that they had single-handedly destroyed about eighteen trucks in a truck park near Iraq's border. The

215

next day recce photos showed they had destroyed a used car lot on the edge of Kuwait City.[9]

After weeks of bombardment, the Coalition ground commander decided that the Iraqi army no longer had the strength to stop a combined ground attack. At 0400 on 24 February, the Coalition's ground divisions invaded Kuwait and Iraq. The ground battle was fast and confused. At 0900, 28 February, when President Bush ordered a cease-fire, the Persian Gulf War ended, except for the withdrawal logistic operation. The short and successful operation caused fewer casualties to U.S. units than most peacetime exercises involving a similar number of units. The total carrier aircraft losses during the forty-two-day air campaign were one F/A-18, one F-14, and three A-6s. There were no A-7 or support airplane losses. The joint Air Force and Navy aircraft loss rate was about one-tenth of the loss rate during the Vietnam War.

From the night the Persian Gulf War started, the Navy's aircraft carriers sustained air combat activities twenty-four hours a day for six weeks. The air wings flew a little over half of their missions at night. The A-6 squadrons in the *Ranger*, the night carrier, flew 85 percent of their missions at night. Carrier night operations took a step toward perfection in this conflict. Aircraft radars, infrared sensors, and laser-guided bombs enabled accurate bombing of fixed installations and improved carrier aviation's ability to find and destroy small tactical targets at night. Although there were still some aircraft losses, the night proved to be a strong element protecting the participants. Clearly, carrier night air combat operations had proven their value to the U.S. Navy.

EPILOGUE

Night and all-weather air combat operations from air-craft carriers have increased dramatically since World War II, when the Navy counted only a handful of specially trained night and all-weather fighter pilots and night bomber crews. Each of the few night pilots of World War II accumulated an average of less than two dozen carrier night landings (the number of landings is a good measure of any carrier pilot's experience) during almost two years of carrier-based night air combat activity. After the war, although the night fighter and attack pilots in the night VC squadrons made 25 to 30 percent of their carrier landings at night during a flying career, the average carrier pilot who started flying in the late 1940s or early 1950s ended his flying career with only 5 to 10 percent of his carrier landings at night.

When the Navy introduced optical landing aids and the angled deck that eliminated the threat of barrier crashes, the intensity of carrier night operations started to increase. A typical pilot, starting in ADs or the first night Banshee squadrons, ended a career with about the same percentage of carrier night landings as the earlier night VC squadron pilots. By 1990, when the Persian Gulf War started, all carrier pilots performed at least 30 percent of their carrier landings at night. Some carrier aviators now approach the end of their flying career with nearly 50 percent of their carrier landings at night.

Through those fifty years, commanders added more night and all-weather operations because flying from carriers became safer. In 1950, in Naval Aviation there were about 54 major accidents per 100,000 flight hours. By 1959, when the angled deck carriers were fully operational, that rate had been halved. Since the late 1980s, the accident rate has hovered around 3 accidents per 100,000 flight hours, dropping one year to almost 1 accident per 100,000 flight hours.[1]

Changes in training and operational procedures as well as improvements in carrier and aircraft systems have helped to improve safety. Keeping all pilots in an air wing night-qualified is now paramount with squadron commanders. Even during periods of austere operational funding, commanding officers stress that the squadron pilots have to maintain proficiency as night flyers despite little time in the air for other types of tactical training. Moreover, commanders have become more concerned about the risk of flying at night or in foul weather. As recently as the middle 1980s, pilots were launched on a simulated surface-search mission at night with the rain so bad that they could not see the *America*'s bow from the waist catapult.[2] A current carrier admiral would probably not authorize flight operations in such severe weather conditions, unless the nation was at war.

At the time this book appears, the Navy is building a fleet around a core of large nuclear-powered aircraft carriers with new systems that will further improve carrier operations at night and in inclement weather. The carriers will operate today's excellent night fighter-bombers, F-14s and F/A-18s, into the twenty-first century. The latest version of the F-14, the F-14D introduced in 1995, has more powerful engines than earlier models and carries up to six 1,000-pound general-purpose bombs or LGBs. These airplanes also carry an infrared and laser sensor system giving the F-14 crews a night and low-visibility surface-search and air-to-ground weapon delivery capability. The latest Tomcat is now a quite capable long-range night attack aircraft. Furthermore, it remains an excellent night and all-weather fighter.

In the 1970s, the Defense Department began work on airplane designs that could evade radar detection by deflecting or absorbing radar signals. The first such "stealthy" airplane was the Air Force's F-117. To make its stealthy design even less detectable, this airplane operated only at night. Following the trend to stealthy airplanes, the Navy will have larger and stealthier F/A-18s in the fleet by 2001. The single-crew F/A-18E's and dual-crew F/A-18F's sensors, instrumentation package, and systems such as the Global Positioning System will make the future

218

Hornets the best night fighter and attack aircraft in the world. Furthermore, the Joint Strike Fighter, a new, stealthy, night-capable fighter-attack airplane, is already on the drawing board. It will further improve the Navy and Marine Corps's twenty-four-hour air combat capacity when it appears later in the twenty-first century.

Modern weapon systems add to the Navy's future night air combat potential. The all-aspect, radar-guided AIM-120 AMRAAM is a long-range air-to-air missile, but it is also the preferred dogfight weapon. Pilots will use it more than their gun, and difficult night intercepts ending within gun range may disappear. New smart bombs such as the Joint Defense Advanced Munition (JDAM) will give the Navy an even more effective night and all-weather potential for attacking land installations.

The night vision goggles that came into use in the late 1980s provide a tremendous improvement in night vision for F-14 and F/A-18 pilots. Pilots report that, when the goggles are available, they fly all night missions using the NVG system. The goggles allow pilots to see almost as well in night conditions with no moon—just starlight—as they do in daylight. Some pilots say that night bombing is no more difficult than day bombing because of the improvement in night vision when using the NVGs.[3]

The slogan "The Night Belongs to Us," adopted during the Persian Gulf War, certainly should be true of future carrier aviation, allowing Navy commanders to expect equal effectiveness from their air wings in day, night, and adverse weather conditions. However, will night air combat operations be as effective as daylight operations? Not quite! During the Vietnam War, Navy operational planners learned that with at least two carriers, flight operations can be conducted continuously night and day. However, there are factors that reduce the maximum number of sorties at night or in foul weather to less than the number flown in clear daylight. These factors are (1) the time interval required between airplanes for safe flight during launch and recovery operations, (2) diminished depth perception of flight deck personnel at night, causing a safety hazard if the pace of operations is not reduced, and (3) in wartime, probable electronic silence and darkened ship conditions, bringing back memories of a minimal number of night sorties from carriers with poor navigation aids, dim flight-deck lights, and poorly lit LSOs. (However, the future night airplanes' electronic systems and night vision goggles promise to allow full exploitation of the carrier's potential even in the darkest of conditions.)

Despite new equipment that eases the pilots' workload, today's carrier pilot's toughest task remains combat at night or in instrument flight conditions.[4] Carrier pilots continue to respect the art of night and all-weather flying. As they sit waiting for the next night catapult launch or prepare to descend from a starlit sky to a landing under a gray cloud layer, Naval Aviators still feel the extra burst of energy needed to get them through the dark sky above a black sea.

Appendix A *Description of Carrier Airplanes*

Propeller-driven Fighters

F4U-2 Corsair. First night fighter. In service 1944–45. Single-seat. Optical gunsight. First AI search radar. Six .50-cal. guns.

F6F-3N/5N Hellcat. In service 1944–49. Single-seat. Optical gunsight. AI search
radar. Six .50-cal. guns; air-to-ground ordnance: 2,000 lb. bombs or rockets.

F4U-5N/NL Corsair. In service 1949–55. Single-seat. Optical gunsight. AI search
radar. Four 20-mm cannon; air-to-ground ordnance: 2,000 lb. bombs or
rockets. F4U-5NL: Winterized with deicers and heaters.

Propeller-driven Bombers

TBM-1 and TBM-3D/E/N Avenger. First night attack airplane. In service 1944–50.
Crew of three. Optical gunsight. First air-to-ground search radar. Two .50-cal.
guns, top gun turret; air-to-ground ordnance: 2,000 lb. bombs or rockets.
TBM-3E/N: improved air-to-ground search radar. TBM-3N: no turret.

AJ-1/2 Savage. First nuclear bomber. In service 1949–59. Crew of three. Air-to-
ground search, targeting radar. First bombing computer. Air-to-ground ordnance: 7,000 lb. nuclear bombs.

AD-3N/4N/NL/5N Skyraider. In service 1950–59. Crew of three. Optical gunsight. Air-to-ground search radar. Four 20-mm cannon; air-to-ground ordnance: 10,000 lb. bombs or rockets. AD-5N: improved air-to-ground radar,
loft bombing system, and nuclear bombs.

Jets

F2H-2N Banshee. First jet night fighter. In service 1951–54. Single-seat. Optical
gunsight. AI search, track radar. Four 20-mm cannon.

F3D-2 Skyknight. In service 1952–55. Crew of two. Optical or radar-aided gunsight. AI search, track, plus tail-warning radar. Four 20-mm cannon.

F2H-3/4 Big Banshee. In service 1954–59. Single-seat. Optical or radar-aided
gunsight. AI, air-to-ground search, and AI track radar (first radar with sta-

bilized antenna). Four 20-mm cannon; air-to-ground ordnance: 3,000 lb. bombs, rockets, or nuclear bombs.

F4D-2 Skyray. In service 1956–62. Single-seat. Optical or radar-aided gunsight. AI search, track radar. Four 20-mm cannon, two AIM-9.

F3H-2/2N/M Demon. In service 1956–64. Single-seat. Optical and radar-aided gunsight. AI search, track radar. Four 20-mm cannon, two AIM-9; air-to-ground ordnance: 6,000 lb. bombs or rockets. F3H-2/2M: added two AIM-7, Sparrow control radar.

A-4C/E/F Skyhawk. In service 1956–73. Single-seat. Optical gunsight. Air-to-ground search radar. Loft bombing system. Two 20-mm cannon; air-to-ground ordnance: 8,000 lb. bombs, rockets, air-to-ground missiles, or nuclear bombs. A-4E/F: analog bombing computer.

A-3A/B Skywarrior. In service 1956–64. Crew of three. Air-to-ground search, targeting radar. Bombing computer. Air-to-ground ordnance: 7,000 lb. conventional or nuclear bombs.

F-8D/E/J/S Crusader. In service 1962–77. Single-seat. Optical or radar-aided gunsight. Infrared (IR) sensor. AI fire-control radar system. Four 20-mm cannon, four AIM-9. F-8E/J/S: air-to-ground ordnance: 4,000 lb. bombs or rockets.

F-4B/J/N/S Phantom II. In service 1960–87. Crew of two. AI search, track, plus Sparrow control radar. IR sensor. Four AIM-9, four AIM-7; air-to-ground ordnance: 10,000 lb. bombs or rockets. F-4J/S: radar replaced with Doppler lookdown radar and improved missile control system. F-4J/N/S: electro-optical (EO) sensor.

A-5A Vigilante. In service 1961–63. Crew of two. Air-to-ground search, targeting radar. Analog bombing system. First INS. Air-to-ground ordnance: 7,000 lb. nuclear bombs.

A-6A/B/C/E Intruder. In service 1963–97. Crew of two. Optical gunsight. First digital weapon system computer. Digital displays. "Pathway in the sky" attitude indicator. INS. Air-to-ground ordnance: 18,000 lb. bombs, rockets, guided missiles, or nuclear bombs. A-6A/B/C: air-to-ground search and separate track radar with AMTI. A-6B used for Standard ARM only. A-6C added first air-to-ground IR/laser system, "TRIM." A-6E: single, improved air-to-ground search, track, and AMTI radar and TRAM turret with FLIR and laser designator.

A-7A/B/E Corsair II. In service 1963–97. Mk-61 20-mm cannon. Air-to-ground ordnance: 11,000 lb. bombs, rockets, guided missiles, or nuclear bombs. A-7A/B: optical gunsight, analog weapons computer, air-to-ground search radar. A-7E: HUD, INS, digital system computer, improved air-to-ground search, track radar. FLIR in later A-7Es.

F-14A/B/D Tomcat. In service 1973–present. Crew of two. HUD, digital displays, digital system computer, digital missile control system. Doppler AI search, track radar. INS. Mk-61 20-mm cannon, four AIM-9, six AIM-54/AIM-7/AIM-120. F-14B/D: improved engines. F-14D added FLIR, laser system, 8,000 lb. bombs or guided air-to-ground weapons.

F/A-18C Hornet. In service 1982–present. Single-seat. HUD, INS, digital displays, digital system computer. AI, air-to-ground search, track radar. Mk-61 20-mm cannon, four AIM-9 and four AIM-7/AIM-120; air-to-ground ordnance: 8,000 lb. bombs, rockets, guided missiles, or nuclear bombs. Improved radar, FLIR, laser designator, and NVG system in later models.

Appendix B   *Performance of Carrier Airplanes*

Name	Engine/Power (HP or lb. thrust)	Weight (lb.)	Radius (NM)	Maximum Speed (Knots/Mach)
F4U-2	R2800/2,000	11,000	450	365
F6F-3N/5N	R2800/2,000	11,000	500	350
F4U-5N/NL	R2800/2,400	12,500	500	370
TBM-1/3D/E/N	R2600/1,700	14,000	600	330
AJ-1/2	One J33/4,600 and two R2800/2,300 ea.	36,000	600	400
AD-3N/4N/NL/5N	R3350/2,800	12,800	550	280
F2H-2N/2B	Two J34/3,250 ea.	15,500	600	500
F3D-2	Two J34/3,250 ea.	20,000	400	500
F2H-3/4	Two J34/3,250 ea.	17,000	600	500
F4D-2	J57/10,000 norm (16,000 mil)	21,000	300	650
F3H-2/2N/M	J71/9,500 norm (14,000 mil)	33,000	400	600
A-4C/E/F	J52/8,000	12,000	400	525
A-3A/B	Two J57/10,000 ea.	38,000	700	525
F-8D/E/J/S	J57/10,000 norm (16,000 mil)	28,000	300	Mach 2.2
F-4B/J/N/S	Two J79/10,000 ea. norm (16,000 ea. mil)	46,000	350	Mach 2.5
A-5A	Two J79/10,000 ea. norm (16,000 ea. mil)	50,000	400	Mach 2.5
A-6A/B/C/E	Two J52/8,500 ea.	28,000	750	525
A-7A/B/E	TF30/11,500	15,000	600	525
F-14A/B/D	Two TF30 or F100/12,000 ea. norm (21,000 ea. mil)	48,000	350	Mach 2.5
F/A-18C	Two F404/15,000 ea. norm (17,000 ea. mil)	24,000	300	Mach 1.8

Notes

Chapter 1. Early Carrier Night
and All-Weather Operations

1. "Naval Aviation," *Aerospace Historian,* Winter 1975.

2. D. V. Gallery, "Postage Stamp Was No Word for It," *Naval Aviation News,* March 1972, 33–34.

3. "Flying the Navy's Wasps at Night," *The Bee-Hive,* January 1932, 7–9.

4. U.S. Atlantic Fleet Aircraft Squadrons, Battle Fleet, *Aircraft Tactics, Development of,* 3 February 1927.

Chapter 2. World War II

1. R. M. Lindsey, ". . . and the Breed of Man Who Brought 'em In," *Naval Aviation News,* November 1967, 16–17.

2. R. E. Harmer, note to author, 4 October 1998.

3. T. F. Caldwell, "An Early Night Air Group," *Foundation,* March 1982, 15–19.

4. M. I. Fox, interview with author, 6 October 1997. The incident was related to Commander Fox by his father-in-law, Capt. R. Wood, USN (Ret.).

5. E. P. Stafford, *The Big E* (New York: Random House, 1962), 260–63.

6. Stafford, *The Big E,* 292–96.

7. R. W. Bruce and C. R. Leonard, *Crommelin's Thunderbirds* (Annapolis, Md.: Naval Institute Press, 1957), 15–16.

8. Department of the Navy, *U.S. Naval Aviation, 1910–1980* (Washington: GPO, 1981).

9. Caldwell, "Early Night Air Group."

10. Ibid.

11. B. Tillman, "The Indispensable Man, the LSO, Part Two," *The Hook,* Winter 1982, 14–23.

12. J. P. Purcell, Jr., interview with author, 6 September 1997. Purcell was a pilot in VF(N)-90 aboard the *Enterprise* during World War II.

13. Bruce and Leonard, *Crommelin's Thunderbirds,* 35, 103.

14. J. W. MacGlashing, *Batmen: Night Air Group 90 in World War II* (St. Paul, Minn.: Phalanx Publishing Co., 1995), 28.

15. Ibid.

16. The Navy has used three words to describe "atomic" weapons: atomic, special, and nuclear. Further usage in this book will be *nuclear.*

Chapter 3. Postwar Transition

1. FAWTUPac Commanding Officer, "Squadron History, Period 1 April 1946 to 30 June 1949, Submission of," letter dated 22 August 1949.

2. D. D. Engen, *Wings and Warriors* (Washington: Smithsonian Institution Press, 1997), 82.

3. R. K. Smyth and R. Good, "A History of VC-4," unpublished, 19 September 1996.

4. R. Morgan, "The Firebirds of VAQ-33," *The Hook,* Winter 1983, 11–21.

5. D. D. Engen, interview with author, 9 July 1997.

6. VC-5 Squadron Histories, 1954–64, Naval Aviation History Center.

7. E. T. Wooldridge, interview with author, 9 July 1997.

8. G. E. Miller, interview with author, 16 July 1997.

9. H. Kiker, interview with author, spring 1997.

10. J. Burden, interview with author, summer 1997.

11. Smyth and Good, "History of VC-4."

12. Miller, interview.

Chapter 4. The Korean War Years

1. Miller, interview. Miller was on Admiral Ewen's staff at that time.

2. H. Ettinger, "Night Hookers, Part II," *The Hook,* Summer 1988, 22–44.

3. R. P. Hallion, *The Naval Air War in Korea* (Baltimore: Nautical and Aviation Publishing Co., 1986), 102.

4. B. Tillman, "Night Hookers, Part IV," *The Hook,* Winter 1988, 12–23.

5. Ettinger, "Night Hookers, Part II."

6. Smyth and Good, "History of VC-4."

7. Tillman, "Night Hookers, Part IV."

8. Smyth and Good, "History of VC-4."

9. R. F. Hunt, notes dated 30 October 1993.

10. G. G. O'Rourke, "Night Hookers, Part III," *The Hook,* Fall 1988, 46–59.

11. The author was the *Boxer*'s officer-of-the-deck during General Quarters that night.

12. Rear Admiral Arnold recounted his experiences to the author after the Korean War.

13. C. E. Hill, Jr., interview with author, 14 August 1997.

14. VC-5 Squadron Histories.

15. VC-6 Squadron Histories, 1954–65, Naval Aviation History Center.

16. Ettinger, "Night Hookers, Part II," and Tillman, "Night Hookers, Part IV."

17. A. A. Schaufelberger, interview with author, 5 September 1997.

18. Smyth and Good, "History of VC-4."

19. R. Lyon, notes dated 15 October 1989. Author also recalls stories of that incident told by Naval Air Training Command instructors in 1954.

20. This and the following incidents are described in Lyon, notes.

Chapter 5. Carrier Night and All-Weather Operations Mature

1. J. R. C. Mitchell, letter to author, 24 July 1997.

2. Smyth and Good, "History of VC-4."

3. Wooldridge, interview.

4. VF-14 Squadron Histories, 1954–57, Naval Aviation History Center.

5. R. C. Mandeville, interview with author, 2 September 1997.

6. Mitchell, letter.

7. Tillman, "Night Hookers, Part IV."

8. Smyth and Good, "History of VC-4."

9. B. Tillman, "Night Hookers, Part V," *The Hook*, Summer 1989, 22–37.

10. "VC-4 Develops Night LSO Suit," *Naval Aviation News*, May 1953

11. Morgan, "Firebirds."

12. J. M. Zacharias, interview with author, 3 September 1997.

Chapter 6. A New Era for Carrier Night and All-Weather Operations

1. Mitchell, letter.

2. Miller, interview.

3. Tillman, "Night Hookers, Part V."

4. R. E. Tucker, interview with author, 22 August 1998.

5. VAH-1 Squadron Histories, 1956–64, Naval Aviation History Center.

6. The author was CTF-60's strike warfare officer in the *Forrestal* during her 1962/63 deployment.

7. J. B. Wilkinson, Jr., interview with author, 27 May 1997.

8. M. E. Patrick, interview with author, 4 June 1997.

9. Mandeville, interview.

10. Schaufelberger, interview.

11. The author was a member of that test team.

12. R. S. Owens, note to author, 11 February 1998.

13. The author was the pilot on that flight.

14. The author flew in the project's night attack phase.

15. The author made the recommendation to Rear Admiral Hyland and observed the incident in VA-83's ready room.

16. W. R. Westerman, interview with author, 3 July 1997.

17. Hill, interview.

18. Mandeville, interview.

19. Zacharias, interview.

Chapter 7. The Vietnam War

1. J. A. Lockard, interview with author, 10 July 1997.

2. Tillman, "Night Hookers, Part V."

3. Mandeville, interview.

4. Zacharias, interview.

5. Mandeville, interview.

6. Wilkinson, interview.

7. Patrick, interview.

8. Zacharias, interview.

9. Schaufelberger, interview.

10. Tillman, "Night Hookers, Part V."

11. Patrick, interview.

12. Tucker, interview.

13. Owens, note.

Chapter 8. The Cold War Continues

1. Fox, interview.

2. S. B. Barnes, interview with author, 2 October 1997.

3. Mandeville, interview.

4. Westerman, interview.

5. Patrick, interview.

6. Lockard, interview.

7. Patrick, interview.

8. Barnes, interview.

9. Fox, interview.

10. D. Taylor, interview with author, 1 October 1997.

11. Fox, interview.

12. J. F. Manning, interview with author, 2 October 1997. As a lieutenant (junior grade), Manning flew F-14s from the *John F. Kennedy* during at least one winter deployment into the Greenland-Iceland-United Kingdom gap.

13. J. Price, interview with author, 15 September 1997.

14. Fox, interview.
15. Owens, note.
16. Barnes, interview.
17. Fox, interview.
18. Tillman, "Night Hookers, Part V."
19. Fox, interview.
20. Price, interview.

Chapter 9. The Persian Gulf War

1. Department of the Navy, *The United States Navy in "Desert Shield," "Desert Storm"* (Washington: GPO, 1991).
2. K. B. Baragar, interview with author, 15 September 1997.
3. M. I. Fox, "Stirring Up a Hornet's Nest," *Foundation*, Fall 1996, 44–57.
4. Taylor, interview.
5. Fox, interview.
6. Barnes, interview.
7. Fox, interview.
8. Barnes, interview. Barnes was operations officer of VA-155.
9. Barnes, interview.

Epilogue

1. Naval Aviation Safety Center, "Accident Statistics," note to author, 15 September 1997.
2. Fox, interview.
3. Taylor, interview.
4. Lockard, interview.

Glossary

Airborne Battlefield Command and Control Center (ABCCC): A large Air Force airplane configured to control strike groups.

Airborne early warning (AEW): Naval mission using airplanes to detect enemy airplanes and missiles at very long range from protected ships and land installations.

Air intercept controller: Radar operator in a ship or shore station qualified to control fighters attempting intercepts, night or day, on enemy aircraft. The successor to the World War II fighter direction officers.

Aircraft carrier designations:

CV: aircraft carrier

CVA: attack aircraft carrier

CVAN: attack aircraft carrier, nuclear-powered

CVB: large aircraft carrier

CVE: escort aircraft carrier

CVL: light aircraft carrier

CVN: aircraft carrier, nuclear-powered

CVS: antisubmarine aircraft carrier

Aircraft carrier air group or wing, squadron, and detachment designations:

CVG: carrier air group (later CVW)

CVG(N): carrier air group specializing in night operations

CVLG: carrier air group for light carriers

CVLG(N): light carrier air group specializing in night operations

CVW: carrier air wing (formerly CVG)

HS: helicopter antisubmarine squadron

VA: attack (later light attack) squadron

VAH: heavy attack squadron

VAM: medium attack squadron

VAN: night attack detachment

VAW: airborne early warning squadron

VB: bombing squadron

VC: composite squadron or air group for escort carriers. Later, squadrons with mixed aircraft types deploying detachments in aircraft carriers for ASW, night fighter, night attack, AEW, or ECM missions.

VF: fighting (later fighter) squadron

VFN: night fighter detachment

VS: scouting (later antisubmarine) squadron

Antisubmarine warfare (ASW): Naval mission using airplanes to detect and destroy submarines.

B-scope: Radar display that presents target range and bearing or range and height in rectangular coordinates. Much harder to interpret than the later PPI scopes.

Bandit: Air contact identified as hostile.

Barrier combat air patrol (BARCAP): Fighter patrols established at a geographical point protecting ships or land installations.

Bogie: Unidentified air contact.

Bullseye: Designated reference point from which airborne enemy fighter locations are reported. Also, center of a target.

Catapult: On an aircraft carrier, a machine designed to propel an airplane with a force sufficient to allow the airplane to reach flying speed.

Combat air patrol (CAP): Fighter patrols established at points relative to a moving ship's or naval force's position.

Combat Information Center (CIC): Ship's compartment in which radar and other sensor information is collected and analyzed and where recommendations for action are made. Originally known as Radar Plot.

Cycle: Time period in which some number of carrier airplanes are airborne. The length of the cycle is determined by the time airplanes can stay in the air. The carrier's flying operation day is divided into cycles in which airplanes launch, fly missions, and recover. On the flight deck at the beginning of a cycle, airplanes launch, then airplanes are recovered. Airplanes are then readied for the next cycle's launch.

Digital display indicator (DDI): Digital aircraft instrument display that can show navigation, flight path, or aircraft systems information.

E-scope: Radar display that presents target range and height in rectangular coordinates.

Field carrier landing practice (FCLP): Practicing carrier landing techniques at a land base. Later called mirror landing practice (MLP).

Fighter direction officer (FDO): Personnel trained to interpret radar scopes and guide airborne fighters to intercept enemy airplanes.

g: Unit representing the force of gravity. Used to measure the force on an accelerating airplane.

Gatling gun: Type of machine gun that achieves a rapid rate of fire by using multiple barrels rotating over a breechblock.

Groove: Segment of the landing pattern flown by carrier aircraft just before reaching the ship's stern.

Horizontal situation indicator (HSI): Aircraft instrument, usually analog, displaying heading and distance to radio navigation stations.

Huffer: A small vehicle with an air compressor used to start aircraft engines.

Identification, friend or foe (IFF): An automatic communication system used to identify known friendly aircraft; those not giving an appropriate response are considered possible hostiles. Modern radar displays provide unique symbols that show the IFF status of all aircraft.

Inertial navigation system (INS): Self-contained navigational system using gyroscopes to provide precise position information to moving ships and aircraft.

Instrument hood: A shield fitting over a pilot's helmet, which, when adjusted, prevents the pilot from seeing outside the cockpit.

Iron sight: An optical gun or bomb sight with no computer or radar system to help the pilot aim the weapon.

Laser designator: A laser generator in an airplane that can place and hold a laser spot on a target.

Laser-guided bomb (LGB): General-purpose bomb fitted with a sensor that detects laser spot reflections from a laser designator on a target.

Laydown: A method of delivering a bomb in which the bomber releases the bomb in level flight at an altitude just above the minimum height required to escape the bomb's fragments.

Link trainer: An early airplane simulator used to teach basic instrument flying procedures.

Night vision goggle system (NVG): An infrared system with which the pilot uses special goggles to increase the ability to see at night. Useful in even low-light conditions.

Officer-in-charge: Officer responsible for selected naval units.

Plan position indicator (PPI): Radar display presenting a true horizontal, or *plan*, view of the terrain or airspace seen by the radar.

Pri-fly: Compartment on an aircraft carrier in which the Air Officer and his assistants control the flight deck operations (including launch and landing of airplanes).

Radar Plot: See Combat Information Center.

Red Crown: Call sign for the ship-based airspace control unit in a Positive Identification Radar Advisory Zone.

Rendezvous: Process of joining two or more aircraft into a formation.

Section: A flight of two aircraft.

Vertical display indicator (VDI): An aircraft instrument displaying flight path information.

Weapon System Trainer (WST): A simulator used for training aircrews in the techniques of flying and using their airplanes.

Welded wing formation: A very tight—almost parade—section tactical formation used at night.

Bibliography

Baragar, K. B. (Cdr., USN). Interview with author, 15 September 1997.

Barnes, S. B. (Cdr., USN). Interview with author, 2 October 1997.

Bruce, R. W., and C. R. Leonard. *Crommelin's Thunderbirds.* Annapolis, Md.: Naval Institute Press, 1997.

Burden, J. (Capt., USN-Ret.). Interview with author, summer 1997.

Cagle, M. *The Sea War in Korea.* Annapolis, Md.: Naval Institute Press, 1957.

Caldwell, T. F. "An Early Night Air Group." [Naval Aviation Museum] *Foundation,* March 1982, 15–19.

Department of the Navy. *Naval Aviation Training and Operational Procedures Standardization (NATOPS) Flight Manual: Navy Model A-6E.* Washington: GPO, 1991.

———. *The United States Navy in "Desert Shield," "Desert Storm."* Washington: GPO, 1991.

———. *U.S. Naval Aviation, 1910–1980.* Washington: GPO, 1981.

———. *U.S. Naval Aviation, 1950–1995.* Washington: GPO, 1996.

Engen, D. D. (Vice Adm., USN-Ret.). Interview with author, 9 July 1997.

———. *Wings and Warriors.* Washington: Smithsonian Institution Press, 1997.

Ettinger, H. "Night Hookers, Part II." *The Hook,* Summer 1988, 22–44.

FAWTUPac Commanding Officer. "Squadron History, Period 1 January 1946 to 31 December 1948, Submission of." Letter dated 31 December 1948.

———. "Squadron History, Period 1 April 1946 to 30 June 1949, Submission of." Letter dated 22 August 1949.

———. "Squadron History, Period 1 July 1949 to 31 December 1949, Submission of." Letter dated 17 March 1950.

"Flying the Navy's Wasps at Night," *The Bee-Hive* [Pratt & Whitney company newsletter], January 1932, 7–9.

Fox, M. I. (Cdr., USN). Interview with author, 6 October 1997.

———. "Stirring Up a Hornet's Nest." *Foundation,* Fall 1996, 44–57.

235

Gallery, D. V. (Rear Adm., USN-Ret.) "Postage Stamp Was No Word for It." *Naval Aviation News*, March 1972, 33–34.

Hallion, R. P. *The Naval Air War in Korea*. Baltimore: Nautical and Aviation Publishing Co., 1986.

Harmer, R. E. (Capt., USN-Ret.). Note to author, 4 October 1998.

Hill, C. E., Jr. (Rear Adm., USN-Ret.). Interview with author, 14 August 1997.

Hunt, R. F. (Capt., USN-Ret.). Notes dated 30 October 1993.

Kiker, H. (Cdr., USN-Ret.). Interview with author, spring 1997.

Lindsey, R. M. ". . . and the Breed of Man Who Brought 'em In." *Naval Aviation News*, November 1967, 16–17.

Lockard, J. A. (Vice Adm., USN). Interview with author, 10 July 1997.

Lyon, R. Notes dated 15 October 1989.

MacGlashing, J. W. *Batmen: Night Air Group 90 in World War II*. St. Paul, Minn.: Phalanx Publishing Co., 1995.

Mandeville, R. C. (Rear Adm., USN-Ret.). Interview with author, 2 September 1997.

Manning, J. F. (Capt., USN-Ret.). Interview with author, 2 October 1997.

Mersky, P. B., and N. Polmar. *The Naval Air War in Vietnam*. Annapolis, Md.: Nautical and Aviation Publishing Co., 1981.

Miller, G. E. (Vice Adm., USN-Ret.). Interview with author, 16 July 1997.

Mitchell, J. R. C. (Capt., USN-Ret.). Letter to author, 24 July 1997.

Morgan, R. "The Firebirds of VAQ-33." *The Hook*, Winter 1983, 11–21.

Morrison, S. E. *History of Naval Operations in World War II*. Vols. 7, 12, 13, and 14. Boston: Little, Brown, & Co., 1960.

"Naval Aviation." *Aerospace Historian*, Winter 1975.

Naval Aviation History Center. "Notes on LSO Equipment." Undated.

Naval Aviation Safety Center. "Accident Statistics." Note to author, 15 September 1997.

O'Rourke, G. G. "Night Hookers, Part III." *The Hook*, Fall 1988, 46–59.

Owens, R. S. (Rear Adm., USN-Ret.). Note to author, 11 February 1998.

Patrick, M. E. (Capt., USN-Ret.). Interview with author, 4 June 1997.

Pawlowski, G. L. *Flat-tops and Fledglings*. New York: Castle Books, 1971.

Price, J. (Lt. Cdr., USN). Interview with author, 15 September 1997.

Purcell, J. P., Jr. Interview with author, 6 September 1997.

Schaufelberger, A. A. (Capt., USN-Ret.). Interview with author, 5 September 1997.

Smyth, R. K., and R. Good. "A History of VC-4." Unpublished, 19 September 1996.

Stafford, E. P. *The Big E*. New York: Random House, 1962.

Tate, J. R. "Moonlighters." *Naval Aviation News*, March 1972, 35–36.

Taylor, D. (Lt. Cdr., USN). Interview with author, 1 October 1997.

Terzibaschtich, S. *Aircraft Carriers of the U.S. Navy.* New York: Mayflower Press, 1980.

Tillman, B. "The Indispensable Man, the LSO, Part Two." *The Hook,* Winter 1982, 14–23.

———. "Night Hookers, Part I, 1942–1945." *The Hook,* Spring 1988, 41–52.

———. "Night Hookers, Part IV." *The Hook,* Winter 1988, 12–23.

———. "Night Hookers, Part V." *The Hook,* Summer 1989, 22–37.

Tucker, R. E. (Capt., USN-Ret.). Interview with author, 22 August 1998.

U.S. Atlantic Fleet Aircraft Squadrons, Battle Fleet. *Aircraft Tactics, Development of.* 3 February 1927.

VAH-1 Squadron Histories, 1956–64. Naval Aviation History Center.

"VC-4 Develops Night LSO Suit." *Naval Aviation News,* May 1953.

VC-5 Squadron Histories, 1954–64. Naval Aviation History Center.

VC-6 Squadron Histories, 1954–65. Naval Aviation History Center.

VF-14 Squadron Histories, 1954–57. Naval Aviation History Center.

Westerman, W. R. (Capt., USN Ret.). Interview with author, 3 July 1997.

Wilkinson, J. B., Jr. (Vice Adm., USN-Ret.). Interview with author, 27 May 1997.

Wooldridge, E. T. (Capt., USN-Ret.). Interview with author, 9 July 1997.

Zacharias, J. M. (Capt., USN-Ret.). Interview with author, 3 September 1997.

Index

About the Author

Charles H. Brown graduated from the U.S. Naval Academy in 1952. After two tours of duty aboard ships during the Korean War, he entered flight training, beginning his eighteen-year career of flying from carriers. Brown's first tour as a naval aviator was in VC-35, the only night-attack unit in the Pacific Fleet at the time. His aviation career also included duty in VX-5, a test and tactics development squadron, and a stint as executive and commanding officer of an A-4 squadron, VA-112, during the Vietnam War.

Captain Brown continued to work in aviation after retiring from the Navy—he completed a second career as an operations analyst for both the U.S. Marine Corps and McDonnell Douglas.